Physiological Ecology
of Forest Production

APPLIED BOTANY: A SERIES OF MONOGRAPHS

SERIES EDITOR: J. F. Sutcliffe, University of Sussex, England

The Growth of Bulbs   *A. R. Rees*
Crop Processes in Controlled Environments   *A. R. Rees, K. E. Cockshull, D. W. Hand and R. G. Hurd*
Control of Growth and Productivity in Plants   *D. Morgan*

APPLIED BOTANY AND CROP SCIENCE

SERIES EDITORS: F. L. Milthorpe, Macquarie University, New South Wales, Australia, and R. W. Snaydon, University of Reading, England

Physiological Ecology of Forest Production   *J. J. Landsberg*

Physiological Ecology of Forest Production

J. J. LANDSBERG

CSIRO, Division of Forest Research, P.O. Box 4008,
Queen Victoria Terrace, A.C.T. 2600,
Australia

1986

ACADEMIC PRESS
Harcourt Brace Jovanovich, Publishers
London   Orlando   New York   San Diego   Austin
Toronto   Sydney   Tokyo   Montreal

ACADEMIC PRESS INC. (LONDON) LTD.
24/28 Oval Road,
London NW1 7DX

United States Edition published by
ACADEMIC PRESS INC.
Orlando, Florida 32887

Copyright © 1986 by
ACADEMIC PRESS INC. (LONDON) LTD.

All rights reserved.
No part of this publication may be reproduced or transmitted in any form or by any means, electronic or mechanical. Including photocopy, recording, or any information storage and retrieval system, without permission in writing from the publisher.

British Library Cataloguing in Publication Data

Landsberg, J. J.
 Physiological ecology of forest production.—
 (Applied botany and crop science)
 1. Forest ecology 2. Trees—Physiology
 3. Plant physiology
 I. Title II. Series
 574.–5′2642 QH541.5.F6
 ISBN 0 12 435965 5

Typeset by Latimer Trend & Company Ltd, Plymouth
and printed in Great Britain by
T.J. Press (Padstow) Ltd, Padstow, Cornwall

Preface

The need for a book about the physiological ecology of forests has been clear for some time. The stimulus that caused me to get down to the task of producing one was provided by Professor Fred Milthorpe, who recognized the need and suggested that I write the book for this series. It was to be a book about the physiology of trees, at the whole-plant level, and the way the growth of trees is influenced by environmental conditions. This is essentially what plant ecology is about: why do particular plants grow where they do, how do they cope with their environment, and how will they respond to change, how productive are they—or could they become? These questions can only be answered if we know how the plants function. We need to understand the physiological processes that govern their growth and how those processes are affected by environmental conditions. This book treats the growth of trees and forests from this point of view, hence its title: "Physiological Ecology of Forest Production".

Although various aspects of tree physiology and forest environments have received considerable attention from research scientists in recent years, there exists no collation of this work, and assessment of its implications for people concerned with forests and forestry. Perhaps as a consequence physiological ecology seems to be largely ignored in university courses for foresters and forest scientists. Agriculture, by contrast, is well served by texts such as *An Introduction to Crop Physiology* (Milthorpe and Moorby, 1974, 1979), which deals with "the way in which the various (physiological) processes are integrated to produce the response shown by whole plants when they are grown as a community in a national environment," and *Plants and Microclimate* by Jones (1983b). Such books can be highly recommended to students and researchers concerned with forests and forestry. This book should not only provide the synthesis of present knowledge of tree physiology, environment and their interactions needed by those concerned with forests, but may also be of value and interest to agriculturalists and plant ecologists.

My objective has been to demonstrate how a quantitative approach to whole-plant physiology, coupled with some knowledge of the physics of plant environments, allows us to analyse the way trees grow, and the way environmental factors and management (e.g. drought, competition, fertilization, thinning) affect growth. Research in forestry has traditionally been empirical—treatments are applied and the results observed, usually over periods of years. Short-term fluctuations in growing conditions, and their effects on growth patterns, are often not observed or cannot be interpreted. The approach offered here—of examining the processes involved and attempting to explain observed growth patterns in terms of those processes—will not solve all problems or eliminate the need for conventional field experiments (these will always remain essential), but it will help in interpreting the results of those experiments, and allow safer extrapolation of those results, more accurate predictions and hence more soundly based recommendations to managers. The emphasis, throughout the book, is on integration. Forests are complex systems and attempts to analyse the way they grow in terms of the processes underlying the growth patterns will only be useful if we recognize that no process is independent; many factors and processes interact and the growth of a stand is the integrated results of these interactions over long periods of time.

The quantitative nature of modern physiological ecology gives rise to a large number of equations; I hope these will not deter foresters and forestry students. Most of the equations are very straightforward and their properties and consequences can be easily explored by inserting appropriate parameter values and doing the calculations on a hand calculator. This "numerical exploration" is an immensely valuable way of developing a feeling for the operation of the processes under consideration and their contributions to variations in growth.

I have not attempted a comprehensive literature coverage but trust that my selection has been fair and provides reasonable coverage of the field.

A great many people have helped me through this task. Mrs June Nicholls typed the whole book in its early draft and, with her usual patience, re-typed most of it after the first set of second thoughts. Karin Munro did a great deal of work on the later version and Anita Gracie typed the figure legends. Vlad Mosmondor drew the figures, and then corrected most of them. To all these people, without whom the book would not have been produced, I am most grateful.

A number of friends and colleagues commented on the drafts of various chapters: my thanks to Bob Pearcy and Bob Robichaux, of the University of California (Davis and Berkeley respectively); to Brian Myers, John Raison, Peter Snowden, Sune Linder (who also helped with references and some data on photosynthesis), Ray Leuning and Alan Brown of CSIRO Division of Forest Research, and to Eric Bachelard of the Department of Forestry, Australian National University. Michael Thorpe (DSIR, New Zealand) and Paul Jarvis (Department of Forestry and Natural Resources, Edinburgh) read and commented on the whole manuscript at the penultimate stage. This was indeed the act of good friends. Both contributed greatly to reducing the effects of my shortcomings. Paul in particular put in an amount of work well beyond the call of duty. I thank them, most sincerely. They cannot be held responsible for any mistakes or omissions.

The title owes much to a Sunday morning discussion with my friend Bruce Sutton of the Department of Agronomy and Horticulture, University of Sydney.

Lastly, it is traditional—and rightly so—that I should thank my family. They have cheerfully tolerated disrupted weekends, unsociable evenings and all manner of other inconveniences while I worked on this book. Their support was, and remains, essential to me.

J. J. Landsberg
Canberra, July 1985

Symbols and Definitions

Standard International units are used throughout the book. Units in this list are given as m, kg, Pa, s, °C, mole or appropriate combinations, but in the text the magnitudes may be changed for convenience, or for simple comparison with published data; e.g. although the units may be given as kg or mole in this table, mg or μmole may be used in the text.

As far as possible I have used standard symbols, or symbols consistent with the literature. I have made extensive use of subscripts to avoid confusion. However, because of the range of the material covered in the book it has not always been possible to use the "normal" symbols for some parameters.

Only the main symbols used are given here. Relatively unimportant symbols used only once are not listed. Symbols are defined in the text when they first occur. Considerable use has been made of subscripts to try to reduce confusion in the meaning of symbols.

$a, b, c,$	are used in many equations to denote empirical constants
A	CO_2 assimilation rate (mol m^{-2} s^{-1} or mg m^{-2} s^{-1})
A_{max}	maximum (light-saturated) photosynthesis rate (mol m^{-2} s^{-1})
A_b	basal (trunk) cross-sectional area (m^2)
A_f	leaf area per tree (m^2)
A_s	sapwood cross-sectional area (m^2)
B_s	bound water fraction in saturated wood
c_p	specific heat of air at constant pressure (J kg^{-1} °C^{-1})
C	capacitance
C_f	composite constant relating foliage mass to plant population
C_{aM}, C_{Mf}, C_{Mb}	canopy, leaf and stem drag coefficients (dimensionless)
C_a	ambient CO_2 concentrations (m^3 m^{-3} or, expressed as partial pressure, Pa)

C_i	internal (leaf) CO_2 concentration ($m^3\ m^{-3}$ or Pa)
C_s	amount of water stored by a canopy (m)
d	zero-plane displacement (wind profile parameter) (m)
d_B	stem diameter (m)
D	vapour pressure deficit (Pa)
$D_s(\theta_s)$	hydraulic diffusivity ($m^2\ s^{-1}$)
e_a	partial pressure of water vapour in air (Pa)
$e_s(T)$	saturated vapour pressure of air (Pa) at temperature T
E	Einstein (1 Einstein = 1 mole of quanta)
E_t	transpiration rate (mass flux of water vapour, $kg\ m^{-2}\ s^{-1}$)
E_l	rate of evaporation of water from wet surfaces in canopies
E_T	amount of water transpired, i.e. the integral of E_t over time
F_{ss}	volume fraction of water in saturated wood
g_{aM}	canopy conductance for momentum ($m\ s^{-1}$)
g_c	canopy conductance for CO_2 into leaves ($m\ s^{-1}$)
g_H, g_V	canopy conductances for heat and water vapour ($m\ s^{-1}$)
g_{ref}	reference (maximum) value of g_s ($m\ s^{-1}$ or $mol\ m^{-2}\ s^{-1}$)
g_s	leaf stomatal conductance ($=1/r_s$, $m\ s^{-1}$ or $mol\ m^{-2}\ s^{-1}$)
H	sensible heat flux ($W\ m^{-2}$)
I	amount of rainfall intercepted by a plant canopy (mm)
J_c, J_f, J_x	volume flow of water to or from storage, foliage and xylem ($m^3\ s^{-1}$)
k	is used with a variety of subscripts to denote various coefficients. The two noted below are among the most important
k_c	carboxylation efficiency ($mol\ m^{-2}\ s^{-1}\ Pa^{-1}$)
k_φ	coefficient in the exponential relationship defining the transmittance of a canopy to short-wave radiation

Symbols and Definitions

K_H, K_M, K_V	turbulent exchange coefficients for heat, momentum and water vapour, respectively
K_c	hydraulic conductivity of wood (m s^{-1})
K_s	hydraulic conductivity of saturated soil (Darcy's Law) (m s^{-1})
L^*	leaf area index (m^2 leaf m^{-2} ground)
L_V	root length per unit soil volume (m m^{-3})
$[m]_i$	mineral nutrient concentration in plant tissue i (g kg^{-1})
$[M]$	average mineral (nutrient) concentration in plants (g kg^{-1}); the symbol for a particular nutrient, e.g. N,P, may be substituted for M
n	is used in several equations as an index (power)
N, N_{max}	mass of nitrogen, maximum possible mass of nitrogen in foliage (kg)
N_s	number of moles of solute per unit mass of water (mol kg^{-1})
p	plant population per unit area (m^{-2})
P	atmospheric pressure (Pa)
P_T	rainfall amount (m)
q_D, q_R	water drainage from the root zone, run-off (m)
Q_i	rate of substrate supply to organ i (kg s^{-1})
Q_{10}	ratio of rate of a process at T and $(T+10)°C$
r	vapour concentration (humidity mixing ratio) (mass of water vapour per unit mass dry air)
r_{aM}	canopy resistance to momentum transfer (s m^{-1})
r_c	canopy resistance to vapour transfer (s m^{-1} or m^2 s^{-1} mol^{-1})
r_{bH}, r_{bV}	leaf boundary layer resistances for heat and water vapour, respectively (s m^{-1} or m^2 s mol^{-1})
r_s	leaf stomatal resistance (s m^{-1} or m^2 s mol^{-1})
R	universal gas constant
R_d	dark respiration rate (kg CO$_2$ s^{-1} or mol CO$_2$ s^{-1})
RH	relative humidity
R_s, R_r, R_x, R_c, R_f	resistances to water movement from soil to roots, through roots to xylem, through xylem to or from storage tissues, and from xylem to evaporating surfaces in leaves (Pa s kg^{-1} or Pa s m^{-3})

s^*	slope of saturated vapour pressure/temperature curve (Pa °C^{-1})
S	total amount of short-wave energy per unit area per day (J m^{-2})
S_c	multiple scattering term in radiation absorption equations
t	time (s)
t_φ	maximum possible hours of bright sunshine (h)
t_d	daylength (h)
T	temperature (°C or K)
T_w, T_d	wet bulb temperature, dew point (°C)
$T_{max}, T_{min}, T_{opt}$	maximum, minimum and optimum temperatures (°C) for plant growth processes
u	wind speed (m s^{-1})
v^*	friction velocity (m s^{-1})
v_o	volume of freely available water in sapwood (m^3)
v_f	volume of water in leaves at full turgor (m^{-3})
V_B, V_f, V_s, V_t	volumes of trunk and branch, leaf, sapwood and any plant tissue respectively (m^3)
w_a	mass of apoplasmic water in tissue (kg)
w_s	mass of symplasmic water in tissue (kg)
W	total mass of a tree (kg)
W_{max}	maximum mass of a tree at a particular population, p (kg)
W_B	stem and branch mass of a tree (kg)
W_f	foliage mass on a tree (kg)
W_r	root mass of a tree (kg)
Y	conversion factor: labile carbohydrate to plant dry weight
z	height, or depth into soil (m)
z_0	roughness parameter (m)
α	albedo, reflection coefficient
α_i	mean sun-leaf angle in canopies
α_p	quantum yield (efficiency) of photosynthesis (mol CO$_2$ per mol quantum)

Symbols and Definitions

β	Bowen ratio
γ	psychrometric constant (0.066 Pa °C^{-1})
γ_r, γ_B	fractions of imported assimilate respired by roots (r) and by stems plus branches (B)
γ_f	foliage loss rate (kg s^{-1})
δ_e	vapour pressure difference between the inside of a leaf and the ambient air (kPa)
Δ	denotes a small difference or interval
ε	ratio of molecular weights of water vapour and dry air ($=0.622$)
ε_φ	energy conversion efficiency of a stand (kg J^{-1})
ε_N	nitrogen productivity of foliage (mass foliage produced per unit mass N)
η_*	aerodynamic parameter ($=u_*/u^2$)
η_B, η_f, η_r	partitioning coefficients defining assimilate allocated to stems, foliage and roots respectively
θ	volumetric water content of plant tissues or soil (m^3 m^{-3})
χ	absolute humidity (vapour concentration, kg m^{-3})
λ	latent heat of vaporization of water (2.47 × 10^6 J kg^{-1})
μ_{ai}	specific activity of component parts of plants
ρ_a	density of moist air (kg m^{-3})
ρ_f	foliage area density in the canopy (m^2 m^{-3})
$\rho_{fs}, \rho_{ds}, \rho_w$	density of wet wood, oven-dry wood and water (kg m^{-3})
σ	Stefan-Boltzmann constant (5.67 × 10^{-8} W m^{-2} K^{-4})
σ_f	specific leaf area (m^2 kg^{-1})
τ	shearing stress (Pa)
φ_{abs}	radiant energy absorbed by a stand (J m^{-2})

φ_L	long-wave radiation irradiance (W m^{-2})
φ_n	flux density of net radiation (W m^{-2})
φ_p	photon flux density (mol photon m^{-2} s^{-1})
φ_s	short-wave irradiance (W m^{-2})
φ_f	radiation flux to foliage (W m^{-2})
Φ	osmotic coefficient
ψ	water potential (Pa)
ψ_{crit}	critical value of leaf water potential, at which stomatal closure is assumed to commence (Pa)
$\psi_c, \psi_f, \psi_s, \psi_x$	water potential of storage tissue, foliage soil and xylem, respectively (Pa)
ψ_p, ψ_π	turgor and osmotic potential (Pa)
ξ_g, ξ_f	proportions of total shortwave radiation transmitted through gaps, and through the trees

Contents

Preface	v
Symbols and Definitions	ix

1 Introduction
1.1 Scope and Objectives — 1
1.2 Outline — 3

2 Process rates and weather
2.1 Basic ideas: rates and states — 7
    2.2.1 *Effects of temperature* — 10
2.2 Processes at different levels — 13
2.3 Weather and its time scales — 20
    2.3.1 *Radiant energy* — 20
    2.3.2 *Air temperature* — 22
    2.3.3 *Air humidity* — 25
    2.3.4 *Wind speed* — 27
    2.3.5 *Precipitation* — 27
2.4 Concluding remarks — 28

3 Stand structure and microclimate
3.1 Stand structure — 32
    3.1.1 *Foliage mass, area and distribution* — 32
    3.1.2 *Leaf dynamics* — 37
    3.1.3 *Stem populations and mass* — 39
3.2 Stand microclimate: energy relationships — 42
    3.2.1 *Canopy energy balance* — 43
    3.2.2 *Partitioning absorbed energy* — 44
    3.2.3 *Radiation penetration into and absorption by canopies* — 47
    3.2.4 *Leaf energy balance* — 50
3.3 Stand microclimate: transfer processes — 52
    3.3.1 *Air temperatures and humidity in forests* — 52
    3.3.2 *Turbulent transfer processes above forests* — 55
    3.3.3 *Transfer inside canopies* — 60
3.4 Transpiration from forest communities — 61
3.5 Concluding remarks — 66

4 The carbon balance of leaves

4.1	Photosynthesis	70
4.2	Stomatal conductance	73
	4.2.1 Effects of irradiance and vapour pressure deficits	75
	4.2.2 Effects of water stress	78
4.3	An empirical model of photosynthesis	79
	4.3.1 Parameter values and their variation	82
4.4	Carbon consumption and export by leaves	84
4.5	Concluding remarks	85

5 The carbon balance of trees

5.1	Calculations of canopy photosynthesis	88
5.2	Respiration	93
5.3	Dry matter partitioning	95
5.4	Root mass and turnover	103
5.5	Concluding remarks	109

6 Nutrient dynamics and tree growth

6.1	Nutrient cycling	112
	6.1.1 The geochemical cycle	113
	6.1.2 The biogeochemical cycle	114
	6.1.3 The "biochemical" cycle	122
6.2	Growth in relation to nutrition	124
6.3	Concluding remarks	129

Appendix to Chapter 6 — 131

7 Water relations

7.1	Scope	133
7.2	Water relations parameters	134
7.3	The rooting volume	137
7.4	Water movement through trees	141
7.5	The hydrological balance	151
	7.5.1 Interception losses	152
	7.5.2 Redistribution of rainfall	154
7.6	Consequences of water stress	156
7.7	Concluding remarks	159

Appendix to Chapter 7 — 161

8 Synthesis

8.1	Models: their characteristics, value and limitations	167
	8.1.1 Detailed physiological models	167
	8.1.2 "Top-down" simple models	170
	8.1.3 Model testing	171

8.2	A "Top-down" model of forest productivity	172
	8.2.1 *Limiting factors; temperature*	174
	8.2.2 *Limiting factors; nutrition*	175
	8.2.3 *Limiting factors; water*	175
	8.2.4 *Final formulation and use*	176
8.3	Concluding remarks	178

References 179

Index 193

1 Introduction

1.1 SCOPE AND OBJECTIVES

Physiology is concerned with the functions and properties of living organisms and ecology with the mutual relations between organisms and their environment. The objective of this book, as the title implies, is to analyse and explain the growth of forest trees in terms of the physiological processes involved and the way these processes are affected by the environmental factors to which the trees are subject.

Forests are communities of trees. They may contain a number of species in varying proportions, possibly dominated by one or two, or they may consist of single species. Single-species forest communities, particularly those maintained by the activities of man in a state of (usually unstable) equilibrium, or slow change, constitute relatively simple ecosystems. Multi-species systems, often with one or more understory layers, are among the most complex ecosystems on earth. Discussion in this book is limited to simple ecosystems, although the processes discussed are, in principle, also applicable to systems of greater complexity.

The book is intended to provide a framework for thinking about and analysing forest growth that will serve the student, the research worker and even the forest manager. It provides no "practical" information about forestry—there are many good textbooks which do that (e.g. Daniel *et al.*, 1979)—and it contains much detail at levels of little apparent interest to practical foresters. However, if a forest manager concerns himself with how trees grow and how the system that he manages functions, he will be in a better position to make rational decisions than the manager who relies only on empirical information and experience—valuable as these may be. I have tried to set the details in the context of the whole system, and to demonstrate the connections between the processes discussed and tree growth.

I believe that, while qualitative understanding is essential, and often precedes quantitative analysis, quantitative analysis must be a major objective of physiological ecology. Many of the processes that contribute to the growth of trees are quite well understood. Their rates depend on external conditions and interactions with other processes going on in the plants. Rates can be measured in relation to external conditions. Changes in the rates of physiological processes may be caused by changes in weather conditions, or the environmental conditions that act on the trees may be changed as a result of artificial treatments, such as thinning or fertilization, imposed by forest managers. Changes in the state of a tree may modify the response of a particular process to a change in conditions. If we can quantify the changes in conditions and can express in quantitative terms the way they affect physiological processes, we will be well on the way to predicting the effects on forest growth and productivity of various events, conditions and changes.

Models. To make quantitative predictions the behaviour of systems must be described in mathematical terms; for this reason extensive use is made of models throughout this book. Models may be conceptual, diagrammatic or, when developed far enough, they may be mathematical. All forms are used here.

Some clarification is necessary with regard to the term "model". In conventional forestry a model is commonly understood to be a statistically derived equation or set of equations, usually based on measurements of forest stands. The model can be used to estimate the amount of timber in stands with the same general characteristics as those from which the model(s) was derived. The projected growth curve of a stand is also, usually, estimated from equations based on the observed growth rates of other, similar stands. For the purpose of this book, however, I will define a model as "a formal and precise statement, or set of statements, embodying our current knowledge or hypotheses about the workings of a particular system and its responses to stimuli". By definition, therefore, all the equations in this book are models.

Models may be written to describe systems at any level of organization (see Chapter 2); the more complex a system the higher the level of organization and the more simplifications will be necessary in attempts to describe the behaviour of the system mathematically. More complex systems also (generally) take longer to reach a new equilibrium after disturbance, or in response to a change in environmental conditions, i.e. they have longer response times than systems at lower organizational levels.

Models may be completely empirical, i.e. based on data obtained by measurements, or mechanistic, i.e. based on the physiological and physical processes that underly the way the system responds to stimuli. Mechanistic models will have greater explanatory power than empirical models. After adequate testing they can also be used with greater safety to extrapolate experimental results and to explore the consequences of conditions outside the range of those encountered in the original experimental work. Models are discussed further in the final chapter where I consider how knowledge about the processes contributing to forest growth can be used to construct mechanistic or explanatory models of forest growth and to guide and constrain the development of empirical models. Even if it is not possible to develop accurate, reliable models, the attempt to formulate them leads to valuable insights into the functioning of the system being modelled. It is also a useful guide to the design of experiments.

The objective of experiments in forests, as in any other system, must be to test statements that have general validity. However, much research on forests and forest productivity is essentially empirical, providing information strictly applicable only to the system and the period in which it was obtained. The range of conditions that may occur over the period of a conventional forestry experiment is enormous, and the chances of the particular combination of conditions, which occurred during an experiment, being repeated is minute. It follows that extrapolation from the results of field experiments, without taking account of the inter-relationships between the physiological and physical factors that caused the observed responses, is a very unsafe procedure. Wherever possible detailed measurements of both environmental conditions and growth responses at various levels (see Chapter 2) should be made in field experiments; the interpretation of these data will be greatly aided by an appropriate framework of ecophysiological theory.

I have, throughout this book, assumed that quantitative and generally deterministic relationships can be applied to processes in trees. Yet trees are living organisms in which diversity and variation are intrinsic properties, so that functional relationships, whether applied to processes or individuals, can only be deterministic in terms of some mean response. There will always be a degree of uncertainty associated with the actual value of a parameter or rate of a process.

Mayr (1983) points out that all biological phenomena have a proximate cause and an evolutionary cause. I am concerned only with proximate causes, which have to do with the way physiological processes respond to immediate stimuli, with the decoding and expression of the genetic programme of individuals. Evolutionary causes have to do with the changes and adaptations of genetic programmes through time. Evolution, adaptation and similar properties of biological organisms are not discussed, but we should not forget that, in the rather simplistic, mechanistic approach used in this book, a great deal of complexity, uncertainty and variation, at every level, has been passed by.

1.2 OUTLINE

The growth of a tree depends on a great many dynamic (non-static) processes, at various levels—from molecular through cellular to organ (leaves, buds, etc.) and the whole plant—each proceeding at a rate determined by environmental conditions, the state of the tree and its genetic make-up. The state of the tree at any time, in terms of its size and condition, is determined by the integral of the interacting rate processes. The rate at which a forest produces biomass is the sum of the growth rates of the individual trees, and the biomass present at any time depends on the number of trees present and the integral, in terms of time, of their growth rates. These ideas are basic to this book and are discussed in Chapter 2.

Because the environment in which physiological processes operate is determined by weather, it is important to appreciate the physical meaning of the various weather parameters; Chapter 2 contains a section giving useful basic information.

The conditions within a forest stand—the microclimate—are determined by interactions between the structure of the stand (height, leaf area, foliage and branch density and distribution) and the prevailing weather. A relatively detailed treatment of stand structure and microclimate is provided in Chapter 3. This gives the background information essential for understanding transpiration and photosynthesis and hence plant water relations and the carbon balance of canopies.

Photosynthesis and the carbon balance of leaves are discussed in Chapter 4. Stomatal conductance, which is important to both photosynthesis and transpiration, is considered in this chapter. Utilizing information provided in Chapters 3 and 4 the carbon balance of canopies and the vital matter of carbohydrate partitioning—including root turnover—are dealt with in Chapter 5.

Chapter 6, concerned with nutrient dynamics and tree growth, covers nutrient cycling, uptake and the nutrient status of trees, and growth in relation to nutrition. The discussion of water relations in Chapter 7 includes basic considerations of forest hydrology—again drawing on Chapter 3—leading to appreciation of the water balance in root zones. It also includes a treatment of water movement into and through trees and the consequences (to growth) of water stress.

The growth of trees—or indeed of any plants—is governed and modified by a great many interacting factors. This book deals with some of those considered to be "first order"—of primary importance to understanding the growth of whole plants in the field. This understanding can and should be put to practical use in forest management. Management involves disturbing or altering a system with the intention of achieving a specified result: the consequences of management actions on forestry are usually guessed from experience or on the basis of empirical, site-specific models—knowledge of ecophysiology can contribute greater certainty about the consequences of disturbing forests. It is possible to develop relatively simple models, based on physiological and physical considerations, which can be used to estimate the

1 Introduction

productivity of any site and evaluate the consequences of planting (or regeneration) at particular spacings, or of thinning, fertilizing or drought. These models will always contain a high degree of empiricism—or need some "calibration"—but they will be much more versatile, flexible and widely applicable than the completely data-based models currently used in forest management. In Chapter 8 I have indicated how such models might be formulated on the basis of the information presented in the book. This chapter includes some discussion of the research needed to improve our knowledge of the physiological ecology of forests and hence our ability to predict forest productivity.

2 Process Rates and Weather

In this book I am concerned with analysing the growth of forest trees in terms of the physiological processes involved. Such analysis must be in terms of the rates of the processes, which depend on external conditions and interactions with other processes going on in the plants. The external conditions are governed by weather.

The ideas presented about rates and states in this chapter, and their mathematical expression, are elementary, but essential to the rest of the book. The idea of considering plant growth as a hierarchical set of processes, with different response times, is fundamental to mechanistic analysis of plant growth, i.e. analysis (or prediction) of observed responses at a particular level (e.g. growth of an organ; photosynthetic rates) in terms of the mechanisms or processes underlying those responses, at the next level down. So the growth of a leaf might be considered in terms of cell division or expansion, or, more grossly, carbon balance. Because the rates of so many processes are governed—or at least strongly affected—by temperature, a subsection of this chapter is concerned with the effects of temperature and serves to illustrate the general approach I have used.

The second half of the chapter provides some basic information on weather variables.

2.1 BASIC IDEAS: RATES AND STATES

Most of the measurements and observations made on plants in the field describe the state of the plant in terms of attributes such as mass (if harvested), height, size or stage of development. The values obtained from such measurements or observations are usually plotted against time. There is a fundamental flaw in this procedure because time is not a driving variable, i.e. the change in state of a plant is not dependent upon time *per se* but upon some condition, associated with time, which causes the change. We should therefore be concerned to evaluate not only the change of state with time, but the factors affecting that change and how they effect it. This may be formally stated as a simple differential equation

$$dy/dt = f(x) \tag{2.1}$$

which indicates that the rate of change of the property y with time t is a function of the variable x. In fact the situation is seldom so simple and we will often be dealing with cases where $dy/dx = f(x_1, x_2, \ldots, x_n)$. The relationships then become extremely complex (usually unmanageable) unless some of the x's can be regarded as constant, or safely neglected over the period of interest. However, this does not alter the argument that follows. The state of the plant at time t can, at least in principle, be determined by integration:

$$y(t) = \int_{t=0}^{t} f(x)\,dt. \tag{2.2}$$

Rate measurement. One of the basic problems in biological research is to determine the form of $f(x)$—to establish the relationships between the rates of change of y and the amount, intensity or concentration of variables (x); i.e. to determine the functional relationships between y and x.

With modern measurement techniques it is now often possible to measure rates directly. For example, using infrared gas analysers the rate of photosynthesis is measured as rate of net-CO_2 uptake and respiration rate as net-CO_2 efflux. Changes in linear dimensions, such as shoot elongation rate or changes in stem diameters, can be measured by sensitive transducers, the outputs of which are continually recorded, and so on. Rates determined by direct measurement are likely to be more accurate than those determined as the difference between the state of the plant at time 1 ($y(t_1)$) and at a later time 2, divided by the time difference:

$$\frac{y(t_2) - y(t_1)}{t_2 - t_1} = \frac{\Delta y}{\Delta t} \tag{2.3}$$

There are normally errors associated with measurements of y and the difference method requires the use of specified time intervals that may or may not be suitable for the system under study. For example, a standard measurement used in evaluating the growth of trees is the diameter of the stem at "breast height" (conventionally taken as 1.3 m). Let us assume that, when measured at time t_1 the average stem diameter ($d_B = y(t_1)$) of a set of trees is 300 mm; a year later ($\Delta t = 1$ yr), $d_B = y(t_2) = 350$ mm, so that the apparent stem diameter growth rate is 50 mm yr^{-1}. If the uncertainty in the measurement of d_B is, say, ± 2 mm (a conservative value) then $\Delta d_B = \Delta y = 50 \pm 4$ mm, and the uncertainty in the growth rate is almost 10%.

Annual measurements provide little information about the effects on growth of conditions during the year; to obtain such information Δt must be reduced. However, if the same measuring technique is used the error in measurement of d_B will remain about the same, so that estimates of, say, monthly growth rates may be of order $(50/12) \pm 4 \approx 100\%$. Clearly then, such measurements are unlikely to be of much value for studying the effects of environmental conditions on growth rates; for this a more sensitive measuring technique is required. (It would also be necessary to ask if stem diameter provides the best measure of response to the stimulus of interest—whether it be changes in water status or the pattern of response to fertilizer application.)

The method of analysis advocated here therefore consists of expressing the rates of processes as functions of the independent variables which drive them. To illustrate this, plant responses to temperature are discussed in the next section. In this, as throughout this book, the (very simple) mathematics used do not include conventional statistics. I have made no attempt to provide information on the mathematical techniques that may be required; an increasing number of books and papers give this sort of information. Some of the more common and easily manageable functional relationships are discussed by Landsberg (1977), while an excellent text on mathematics for biologists is that of Batschelet (1979).

Role of statistical analysis. Conventional statistics, including statistical design, appropriate replication, analysis of variance, regression and the rest, are an essential part of research in fields such as forestry and agronomy. These techniques are of great value in comparing the effects of treatments upon the state of a system at any time, in evaluating the probability that differences are real, or fitting constants and coefficients to equations and testing whether the values differ between sets of data. Statistical techniques are also important in evaluating the variance associated with sampling, which leads to proper sampling procedure—always essential in research involving measurements in populations.

Multivariate analysis is a technique that may be valuable in hypothesis formulation or in testing the relative importance of particular variables, but it is not a substitute for knowledge about the mechanisms and inter-relationships of a system. Statistical analysis cannot provide any more knowledge about a system than is contained in the data, and even the most cunning manipulation will not extract information from collections of data that are not suited to the testing of a hypothesis. Statistical methods are often used to test the so-called "null hypothesis"—that treatments are not different. This may be trivial and whether rejected or not nothing is learned about the system under study. For example, suppose we measure some property of a group of trees, such as stem diameter. We then apply fertilizer to half the group and, after a suitable interval, return and re-measure the trees. The null hypothesis is that application of fertilizer has no influence on the growth rate of the trees, and hence their size at the end of the time interval. Statistical analysis of the measurements may show that the null hypothesis must be rejected—the probability that the differences in growth between the fertilized and unfertilized trees are not real is so small as to be unacceptable. The differences are therefore real and can be attributed to fertilization. This may be useful information to a manager, but to a biologist it is a trivial result, merely confirming what is already known.

Of much more interest are questions such as: would the magnitude of the difference be the same on different soil types, with a different rainfall pattern, with trees of a different age, spacing, etc.? These questions could be answered by more experiments, but each one would merely provide another example of the phenomenon. We need to understand the functioning of the system in terms of the processes involved—the rate of nutrient uptake under specified conditions and growth rates in relation to the nutrient concentration in the tissues. These matters will be discussed in some detail in Chapter 6; here the example is used to illustrate why we are concerned with functional relationships and not with conventional field experimentation.

In physiological modelling and experimentation we usually assume that we know which physiological processes are most important in plant growth and what factors affect them. For example, there seems little point at this stage in setting up experiments to test whether light (photon flux) is more important than temperature in determining the rate of photosynthesis; the effects of the two are thoroughly documented. Similarly, it hardly seems justifiable to expend much effort in time-series analysis of diurnal changes in transpiration rate, tissue water potential and stem radius in trees to demonstrate that these can be described by a simple resistance–capacitance model. Such models have already been proposed, formulated, tested and found to explain adequately the observed variations (see Chapter 7). The problem, in many cases, is to evaluate their relevance to trees and obtain appropriate parameter values for the functional relationships between processes and the variables which drive them.

2.1.1 Effects of Temperature

The rates of many plant processes are affected by temperature, so responses to temperature provide useful illustrations of rate dependence and functional relationships.

The effects of temperature on plant growth are largely attributable to its effects on the activity of enzyme systems: a rise in temperature increases the kinetic energy of substrate and enzyme molecules so that more frequent collisions occur and the measured reaction rate becomes greater. A rise in temperature also increases the number of molecules with sufficiently high kinetic energy for the reaction(s) to occur, but if temperatures become excessive there is rapid denaturation of enzymes. Different temperatures therefore lead to quantitative differences in plant growth because physiological processes proceed at different rates in response to temperature. In common with all other environmental factors affecting plant growth, temperature effects can never be usefully considered in isolation. They vary with age of the plant or plant part, plant water and nutritional status, temperature prehistory and radiant energy levels.

Temperature effects are reflected in easily observable processes such as the stage of bud and floral development, leaf emergence rates and stem elongation rates. The usual form of the temperature response curve is shown in Fig. 2.1; this can be described by an equation given by Landsberg (1977). It can also be approximated by a set of straight lines as indicated on the figure.

Figure 2.1 represents the rate at which a process or complex of processes might respond to temperature. In general terms it is described by equation (2.1), now written

$$dy/dt = f(T)$$

with $f(T)$ being the equation describing the curve. The results of many experiments indicate that responses to temperature are exponential in the range $T < T_{opt.1}$, i.e. $f(T) = y_0 \exp[b(T - T_{min})]$, where b is a constant. However, the linear approximation will usually serve well enough for empirical description, i.e. if $T_{min} < T < T_{opt.1}$, $f(T) = b(T - T_{min})$. If $T < T_{min}$, $f(T) = 0$.

Biologists have long recognized that the rate of plant development depends on temperature, but analysis of the relationships has been complicated by the difficulty of expressing developmental stages in quantitative terms. This has led to analysis in terms of "thermal time" or "heat units", which have no real physiological meaning but which can be rationalized and are soundly based as long as the linear approximation holds. Developmental processes usually involve a change of state or condition. This change may be marked by a "point event", e.g. the opening of a flower, that can be easily observed. By beginning the thermal time calculations from some arbitrarily chosen time or observed state (y_1 at time t_1), and assuming that the response to temperature is linear, we can calculate the time taken to reach some new observed state y_2 at time t_2.

Retaining the constraint that $f(T) = 0$ when $T < T_{min}$, the equations involved are

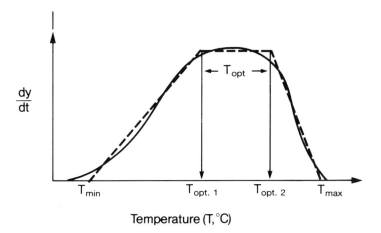

Fig. 2.1. Generalized temperature response curve for a process or complex of processes. The shape of the curve and the positions of the cardinal points (T_{min}, T_{opt}, T_{max}) may vary with species, the process concerned or even with seasons. As a general guide we would expect T_{min} for temperate deciduous trees to be about 0°C, with the middle of the T_{opt} range about 25°C. For tropical trees these values may be 10°C more. The first part of the curve would be approximated by an exponential equation. For purely empirical purposes the whole curve can be approximated by three straight lines (broken) (see also discussion in Chapter 8).

$$\frac{y_2 - y_1}{t_2 - t_1} = \frac{\Delta y}{\Delta t} = b(T - T_{min}) \qquad T_{min} \leqslant T \leqslant T_{opt} \qquad (2.4)$$

hence

$$y(t) = \sum_{i=1}^{n} b(T_i - T_{min}) \Delta t_i = b(T_i - T_{min})(t_n - t_1) \qquad (2.5)$$

where $y(t)$ is the state of the plant at the time t, T is the average temperature of the ith interval, T_{min} is the base or limiting temperature (see Fig. 2.1), b the slope of the relationship and Δt_i is the time interval in days. Equation (2.5) is a finite-difference form of equation (2.2). The units on the right-hand side of equation (2.5) are (temperature × time), i.e. "thermal time"—probably the preferable term. This may be regarded as a form of biological time, since the time taken for a plant to progress from state y_1 to state y_2, other factors being constant or non-limiting, depends on temperature.

There are many examples from crop physiology where the rate of leaf appearance has been shown to be linearly related to temperature so that equations (2.4) and (2.5) apply, and the slope b may be found by plotting leaf number against $(T_i - T_{min}) \Delta t$. Numerous other physiological processes, at various levels, are also temperature dependent; an important example considered later (Chapter 5) is respiration. Cannell and Smith (1983) used a thermal time model to predict the date of bud burst in Sitka spruce (*Picea sitchensis*). They found that the model was improved by introducing a "chill days" factor, which accounted for the influence of low temperatures in breaking dormancy—the longer the buds are exposed to chilling, the more rapidly they will grow when subsequently exposed to warmth. Landsberg (1974) used a similar model to analyse time to bud burst in apples.

The model illustrated in Fig. 2.1 is used in Chapter 8, where the effects of temperature on dry matter production over periods of weeks are considered in the context of an overall, simple model of forest productivity.

2.2 PROCESSES AT DIFFERENT LEVELS: RESPONSE TIMES

The previous section dealt with the changes caused in plants by the effects of temperature on the rates of physiological processes. These changes can be simply observed or measured at the level of organs or the whole plant but at lower levels (cellular, biochemical) more complex techniques are required to identify them. At the higher levels changes in the state or condition of plants are the end result of many physiological processes; the number of processes involved in any particular change decreases at the lower levels of organization. In this section we consider the important question of organizational levels and their response times in more detail.

Molecular and biochemical processes. Physiologists concerned with molecular or biochemical processes may work with species that provide good examples of particular processes, and with easily isolated tissues free from compounds that interfere with chemical analysis. The material may also have to be grown under standard, carefully controlled, conditions. Research at such levels may seem far removed from, and irrelevant to, the complexities of plant growth in the field, but it provides essential information about the basic processes and mechanisms by which plants grow. At the lowest organizational levels all plant growth processes come down to operational units. These units react and interact across their boundaries at rates depending on temperature, enzyme amounts and substrate concentrations, transport systems and balances between compounds in the same or different tissues.

Studies at molecular and biochemical levels are essential, but by their very nature they must be on simplified systems or units within complex (higher order) systems, which can be considered independent of their boundary interactions. We must also recognize that understanding a particular process at molecular and biochemical levels, or identification of different processses in different species, does not necessarily explain observed differences in plant performance or adaptation to particular conditions. Such explanations may rest on a biochemical pathway or system but will usually involve high-level, more complex processes at the level of organs or individuals or stands and their responses to stimuli.

Range of states and processes. The states and processes that describe the condition and functioning of plants and plant communities can be considered as a hierarchical continuum. For convenience this can be arbitrarily divided into groups, or levels, based on the response time of the process concerned (de Wit, 1970, Osmond *et al.*, 1980). Figure 2.2 provides a summary of the states, processes and time scales involved.

Process (Level)	Biochemistry		Physiology		Ecology	
Organ	Cell	Tissues	Organs	Plants	Community	Ecosystem
Response time (s)	$1-10^2$	10^2-10^4	10^5-10^6	10^6-10^7	10^7-10^8	10^8-10^{10}
Time scale	seconds	hours	weeks	months	years	years

Fig. 2.2. Diagrammatic representation of states and processes, and their response times. (Based on Osmond *et al.*, 1980.)

The response time of a process is the time required for a change, in response to a stimulus, from one steady state (or rate) to another. For example, in a laboratory experiment we may, when studying the rate of photosynthesis in relation to photon-flux density (φ_p, see 2.3), use a step change in φ_p to alter the rate of photosynthesis. We would then expect to observe an asymptotic response to the change, taking anywhere between 5 and 60 min (depending on species, state of the plant, etc.) to reach a steady

2 Process Rates and Weather

rate. However, photon-flux density in the field may be changing continuously, so that the system never approaches a steady rate. In fact steady states or rates are unusual in biology and it is not usually useful to try to apply the above definition of response time rigorously. Quasi-steady states, where rates of change are slow or there are continuous fluctuations about a mean value, are more common. The concept of processes at different levels, and their significance for the growth of trees, is illustrated by considering them in the context of models.

Modelling in terms of processes at different levels. De Wit (1970) suggested that it is not practical to attempt to model a system across more than two levels of organization—that is to attempt to explain an observation in terms of processes more than one level down. This is generally true for numerical modelling since, for processes with response time differences of say 100, then to move from state y_1 to state y_2 requires about 100 calculations. To illustrate, we will consider briefly the calculations which might be involved in estimating the daily carbon balances of plants (see Chapter 5 for more detail) and the process of water movement through plants (see Chapter 7), which is a major factor affecting their water status. Water status, like nutrition, is a condition influencing the functioning of the growth processes.

Leaf photosynthetic rates depend non-linearly on photon-flux density, therefore calculations of photosynthesis using values of photon flux averaged over periods of the order of, say, an hour, would almost certainly be biased. CO_2 uptake rates respond rapidly to changes in photon flux (see Fig. 2.3) so photosynthetic rates can be considered constant over any period for which photon flux can be considered constant. This will vary with conditions such as cloudiness, but using 5 min as an averaging period would seldom lead to significant bias. This interval may well be used in detailed calculations of assimilate production by a plant stand, with the net CO_2 uptake being accumulated over a whole day. Because of much slower response times, and less rapid variation in the independent variable—in this case temperature—carbon losses by respiration from stems and branches may only need to be calculated hourly. The carbon balance of the whole plant, or canopy, would be updated daily.

Therefore, over a day with a 12 h photoperiod, we might have about 150 photosynthesis calculations and 24 respiration calculations to update the state of the plant in terms of dry matter and its distribution. An alternative approach would be to divide photon-flux density into, say, 10 levels, take the proportion of the photoperiod spent in each level and then do only one set of calculations for each flux density level. Such calculations would be necessary

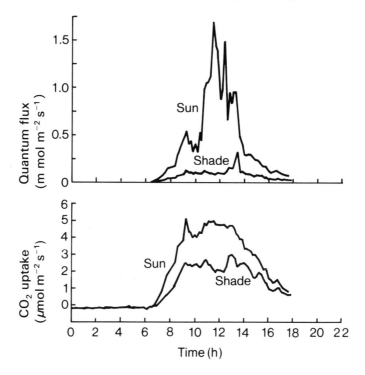

Fig. 2.3. CO_2-uptake by mountain beech (*Nothofagus solandri*) foliage: process at the plant tissue level. Response times of order 10^1–10^2 s. Note the very high correlation with rapid changes in quantum flux (φ_p) (see also Chapter 4). (From Benecke and Nordmeyer (1982), with permission.)

to analyse how leaf photosynthesis might influence plant productivity.

Leaf water status (defined by the leaf water potential, ψ_f) depends on the osmotic potential of the leaf cells and the rate of water movement through the plant, the resistances in the flow pathways, and soil water potential, ψ_s. Rate of flow of water through the plant is obtained from the transpiration rate, which under natural conditions is unlikely to be steady for more than 5–10 min. Shading experiments show that ψ_f responds within minutes to changes in transpiration rate. We may, however, take soil water potential in the root zone as effectively constant for at least an hour. (This is probably true for the bulk soil, but may be a poor assumption for the rhizosphere—the soil in immediate contact with the roots.) Therefore, to calculate the diurnal pattern of leaf water potential, it would probably be necessary to calculate transpiration rate—and hence ψ_f—over 5 min periods, taking ψ_s as constant. The hourly average value of ψ_f could be obtained from the 5 min values. At the end of the hour, soil water content and hence ψ_s can be updated.

Conceptually, of course, it is possible to cross several organizational levels, and low-level processes are frequently invoked as explanations for observations at a higher level. As an example, we may observe that different plant species, or even different provenances within the same species, have different capacities to take up nutrients from soil—reflected in different tissue nutrient concentrations. This might be explained in terms of differences in the absorbing properties of fine roots, identified in careful laboratory studies. The process would have a response time of minutes. This apparently reasonable explanation could, conceptually, be examined at the level of the whole plant by calculations of nutrient movement through soil to roots along concentration gradients caused by uptake, transfer in the plants, and the dynamics of roots. In practice, to write such a model in detail would be futile because the uncertainties at each level are so great that calculations in one part of the model that used results derived from another part would be almost meaningless. For example, calculations of total nutrient uptake by roots must use estimates of effective root length and root length density. Changes in these depend on the carbon balance of the plant and assimilate allocation to its component parts (see Chapter 5), so that in a complete model it would be necessary to incorporate the feed-backs between nutrient uptake and carbohydrate production and allocation. There would be so many parameters with uncertain values involved that almost any answer could be obtained. More realistically, investigations of the causes of different tissue nutrient concentrations might involve careful experimental work on the properties of roots in relation to nutrient uptake (including determination of time constants). They would also involve evaluation of the consequences for growth of differences in (above-ground) tissue nutrient concentrations. We might then use relatively simple models to evaluate the consequences of various possible system states.

In this book attention is confined to processes from the leaf level upwards (to the forest community). Figures 2.3–2.5 show examples of processes at the levels with which we are concerned and illustrate their response times.

Examples of processes at different levels. Figure 2.3 shows the time course of CO_2-uptake by mountain beech (*Nothofagus solandri*). The data were

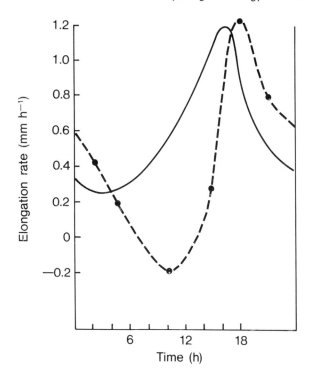

Fig. 2.4. Shoot elongation rates in *Pinus radiata* (broken curve) and *Eucalyptus regnans* (full curve). Response times are of order 10^3–10^4 s. It seems likely that the main factor affecting these rates was changes in water potential (probably turgor potential, ψ_p; see Chapter 7). Note that the greatest elongation rates occurred in the late afternoon, when ψ_p would be increasing as water moved into tissues and labile carbohydrates would be available from photosynthesis during the day (redrawn from Cremer, 1976).

collected in the field using a portable assimilation chamber and illustrate the tight coupling (response times of 10^1–10^2 s) between net photosynthesis and rapid fluctuations in quantum flux (photon-flux density). The differences between rates of CO_2-uptake by leaves in the shade and the sun are not as large as the differences in quantum flux because of the non-linearity of the photosynthesis–light response curve (see Chapter 4). For the same reason the peak sun-values of quantum flux do not cause corresponding peaks in photosynthesis of leaves in the sun; the diagrams indicate that the maximum rate of sun–leaf photosynthesis (4.5 µmol m^{-2} s^{-1}) is reached at a quantum flux of about 0.5 mmol m^{-2} s^{-1} (i.e. about 500 µE m^{-2} s^{-1}).

2 Process Rates and Weather

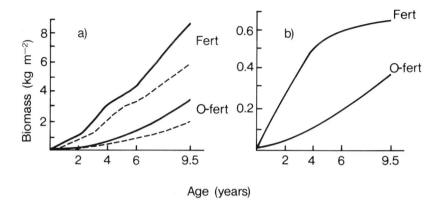

Fig. 2.5. Time course of biomass accumulation in *Eucalyptus globulus*. Response times are of order 10^7–10^8 s. (a) Full curves are total above-ground biomass, broken curves are stemwood biomass. "Fert" denotes heavy fertilization; "0-fert", no fertilization. The ratio (stemwood biomass:above-ground biomass) in the fertilized treatment changed from 0.4 at age 2 to 0.7 at age 9.5 years. (b) Leaf biomass for the same two treatments. The ratio of leaf to total biomass in the fertilized treatment changed from about 0.3 at 2 years to less than 0.1 at 9.5 years (redrawn from Cromer and Williams, 1982).

The curves in Fig. 2.4 illustrate the time course of the elongation rates of shoots of young *Pinus radiata* and *Eucalyptus regnans*. The fact that the curves are out of phase with the time course of irradiance reflects the time needed by the plants to transport the products of photosynthesis to the growing points and there convert them into tissue. The length changes measured almost certainly reflect changes in cell size caused by turgor (see Chapter 7) as well as the response time needed for assimilate conversion. Analysis of such curves might involve calculation of the diurnal course of assimilate production and translocation. Some attempt might also be made to quantify factors such as the rates of cell division in apical meristems as a function of rates of assimilate supply.

A major effect of fertilization on trees is the increase in leaf area and the life span of leaves, resulting in more light interception and hence more photosynthesis. These effects are reflected over periods of years (10^7–10^8 s) in the curves in Fig. 2.5, which show the time course of biomass accumulation in *Eucalyptus globulus* in fertilized and unfertilized plots. Improved fertility caused a rapid increase in leaf biomass which, in the fertilized plots, was greater after about three years than in the unfertilized plots after more than nine years. Useful models of growth patterns and productivity on the time scales of Fig. 2.5 could be written, with total assimilation over intervals of about a week (10^5–10^6 s) calculated from models such as those presented in Chapter 5. Alternatively, empirical relationships between dry matter production and absorbed radiant energy (see Chapter 8) could be used.

2.3 WEATHER AND ITS TIME SCALES

Since weather conditions govern the rates of most processes involved in the growth of trees, and consequently the state of the trees at any time, it is important to understand the physical significance of the terms widely—and often loosely—used to describe weather. We must also recognize that, in any analysis of plant growth in relation to the weather, the interval over which the weather variable of interest is measured or averaged must be appropriate to the response time of the plant process under study. Processes with rapid response times must be analysed in terms of short period averages of the appropriate environmental conditions. Similarly, processes with long response times would be analysed in terms of a suitably averaged variable; e.g. the total amount of biomass accumulated over a growing season might be analysed in terms of total radiant energy received by the forest over the season. Some examples pertinent to this point are given in the following sections.

2.3.1 Radiant Energy

Short-wave radiant energy from the sun (φ_s) is normally measured on a horizontal surface and is the basic energy source driving plant growth, and hence all biological processes. Solar energy reaches the earth, as either direct or diffuse radiation, in wavebands between about 300 and 3000 nanometres (nm). A proportion is reflected and the energy absorbed by the surface is dissipated as latent (λE) or sensible (H) heat (see Chapter 3). The visible part of solar energy (light energy) is in the band from about 370 nm (blue light) to 700 nm (red light). This is often termed "photosynthetically active radiation" (PAR).

When considering a process such as transpiration or the energy balance of bodies, we are concerned with the whole spectrum, including long-wave radiation, and use appropriate units (W m^{-2}, where 1 W = 1 J s^{-1}). When considering photochemical processes such as photosynthesis, the unit now used is the mole (1 mole of quanta = 1 Einstein (E) where a quantum is the energy in a photon, the fundamental particle of radiation). PAR may also be expressed as energy-flux density, but photon-flux density has the great advantage that photosynthetic efficiency can be expressed in terms of the number of moles of CO_2 fixed per mole of photons, i.e. per mole equivalent of quanta (photons) in the visible wavebands in which radiation will activate photosynthesis. Photosynthetically active radiation will subsequently be denoted φ_p, with units $\mu mol\, m^{-2}\, s^{-1}$. (Note, for comparison with many published data, that μmol and μEinstein (μE) are numerically equal.) About half of the incoming short-wave radiation is in the photosynthetically active band. It is not strictly correct to convert directly from energy units to photon flux, because the energy of a photon depends on its wavelength. However, in normal sunlight 1 J of PAR contains about 4.6 μmol, i.e. for solar radiation $1000\, W\, m^{-2} \approx 2200\, \mu mol\, m^{-2}\, s^{-1}$. For a comprehensive account of the basic physics of radiation, see Monteith (1973) and Campbell (1977, 1981), and for an outline of the physics and physiological background of photosynthetically active radiation, see McCree (1981). Subsequent discussion in this section is confined to consideration of φ_s.

The energy from the sun reaching a horizontal surface depends on latitude, season and cloud cover. In temperate regions, summer maximum values of φ_s are likely to be around $800-900\, W\, m^{-2}$. In semi-arid regions values of over $1000\, W\, m^{-2}$ are common.

Under cloudy conditions φ_s may easily vary by a factor of 5 over periods of a few minutes, but on cloudless days it may be assumed effectively constant for periods of about half an hour near the middle of the day when variation is least. On a clear day, the time course of φ_s can be approximated by

$$\varphi_s(t) = \varphi_{s\,max} \cos(\pi t / t_d) \qquad (2.6)$$

where t_d is day length (hours), t is time in hours from midday and $-t_d/2 < t < t_d/2$; $\varphi_{s\,max}$ is maximum radiation intensity. Given an average value for $\varphi_{s\,max}$, equation (2.6) can be used to estimate the average time course of φ_s over, say, a period of a week without gross inaccuracy, although if used for any particular day the correspondence between $\varphi_s(t)$ (estimated) and $\varphi_s(t)$ (measured) over any specific short period may be poor.

The integral of equation (2.6) for one day is

$$S = \int_{-t_d/2}^{t_d/2} \varphi_s\, dt = \frac{2 t_d \varphi_{s\,max}}{\pi} \qquad (2.7)$$

from which the total daily energy income S can be estimated.

Total daily energy income ranges from about $30\, MJ\, m^{-2}\, day^{-1}$ in semi-arid areas to virtually zero in winter at high latitudes. At any location mean daily values of S for each month vary from year to year by only about ±10%.

Because good radiation measurements are difficult to make and because the fundamental importance of radiant energy to all branches of biology has only recently been recognized by those responsible for meteorological networks, values of S are often not available for forested areas. However, for periods of the order of a week or more, S can be estimated from the Angstrøm relationship with sunshine hours, measured by a Campbell–Stokes recorder:

$$S = S_0(a + bn_\varphi/t_\varphi) \tag{2.8}$$

where S_0 is average value of S at the outside of the atmosphere, n_φ is actual hours of bright sunshine and t_φ maximum possible sunshine hours (obtainable from tables). Values of a and b vary with latitude and time of year, but values of $a = 0.23$ and $b = 0.5$ will give good estimates in most areas (see Stigter (1980) for a discussion).

Irradiance is the flux of energy per unit area of a plane surface (Bell and Rose, 1981). Clearly the irradiance incident on sloping and horizontal surfaces will differ. The ratio (irradiance on sloping/irradiance on horizontal) will depend on the slope, latitude and azimuth of the surface, the zenith angle and azimuth of the sun and the proportion of φ_s which is in the direct beam (directional) or is diffuse—in effect coming from all directions. Sellers (1965) gives the equations for calculating direct beam irradiance to sloping surfaces, for any given sun position, while Oke (1978) provides a useful discussion on the effects of slope on irradiance, together with measured values for a number of sites. Obviously the effects of sloping surfaces on irradiance are reduced if much of the energy is in diffuse form. Diffuse radiation is also more effective (per unit flux) for photosynthesis than direct beam. The ratio of energy in the direct beam and diffuse components depends on factors such as cloudiness and the amount of particulate matter in the atmosphere. If the sky is completely overcast all radiation is diffuse, but the opposite does not apply—even under completely clear skies at least 10% of radiation will be diffuse.

2.3.2 Air Temperature

Measurements with very fine, fast-response sensors, show that air temperature may fluctuate rapidly, varying as much as several degrees about the mean. Such fluctuations are usually associated with unstable atmospheric conditions (heat rising from the ground). Short-term fluctuations may be of

2 Process Rates and Weather

concern in relation to detailed studies of leaf energy balance, especially with small leaves which are more closely coupled to air temperature (see Fig. 2.6).

However, despite its rapid short-term fluctuations, air temperature can generally be assumed constant for periods up to half an hour to an hour. Measured with the usual rather massive thermometers, the rapid fluctuations will not be registered and it is doubtful if they are of any significance to plants, even those with leaves closely coupled to air temperatures.

Plant growth processes with response times of the order of hours (e.g. cell division) would be expected to respond to average temperatures over similar periods. They might, therefore, reasonably be analysed in relation to average daily temperatures, usually calculated as the mean of daily maximum and

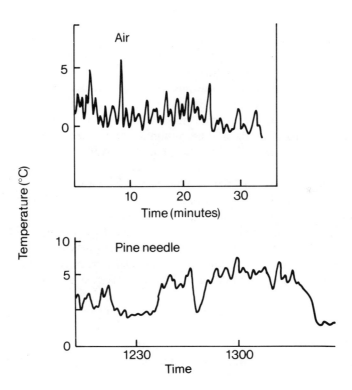

Fig. 2.6. Top. Fluctuations in air temperature, measured by an unshielded thermocouple at noon with variable cloud. Bottom. Temperature variation in a pine needle measured with a very fine thermocouple. There was variable cloud and very little wind. Both sets of measurements reflect the rapidly changing energy balance of bodies of small mass closely coupled to the environment (redrawn from Christersson and Sandstedt, 1978).

minimum. The daily temperature cycle is usually approximately sinusoidal and the temperature at any time ($T(t)$) during the day (24 h) can be estimated from

$$T(t) = \bar{T} + \left(\frac{T_{max} - T_{min}}{2}\right) \cos\left(\frac{2\pi(t - t_{max})}{24}\right) \quad (2.9)$$

where $\bar{T}$ is average daily temperature calculated as $(T_{max} + T_{min})/2$, where t is the time in hours and t_{max} is the time when $T(t) = T_{max}$.

Equation (2.9) might be used to estimate the diurnal course of temperature as a basis for analysing processes such as respiration or transpiration in situations where T_{max} and T_{min} are known. The time t_{max} can be estimated from quantitative knowledge of the local temperature cycle. (Note that temperatures inside forest canopies are not necessarily the same as those measured in standard meteorological enclosures; see Chapter 3.)

For most purposes there seems little reason to depart from using average daily temperatures over a selected time period as the basis for analysing plant growth. However, maximum and minimum temperatures are of interest in themselves because they may cause damage to plants. Excessively high temperatures may inactivate enzyme systems while low temperatures may cause cold or frost damage. The minimum temperature that plants can tolerate varies enormously. Tropical plants will stop growing and may be damaged by temperatures as high as 10°C while trees growing in cold climates may be able to tolerate temperatures as low as −40°C without serious damage. This tolerance may be season specific: very low temperatures in spring, for example, may cause serious damage to buds and growing points in plants which could tolerate much lower temperatures in winter. Therefore, in analysing the climatic range to which a species or provenance is adapted, it is important to examine not only its growth patterns in relation to average temperature but also its ability to withstand extremes at various stages of growth.

Soil temperatures affect plant growth rates, nutrient and water uptake. They vary widely and rapidly near the soil surface, with the amplitude of the diurnal temperature wave decreasing with depth. Soil temperature variation has been extensively and rigorously analysed (see, e.g., Campbell, 1977)—the main problem in applying the analyses is usually the lack of information on the thermal conductivity of soils, although estimates can be made based on soil composition and water content (see de Vries, 1958).

It is not easy to quantify the effects of soil temperatures on growth, particularly in temperate and hot regions. Short-term experiments with seedlings, usually using constant soil temperatures, show that root growth is, as would be expected, directly related to temperature. Work with herbaceous

2 Process Rates and Weather

plants (e.g. Davidson, 1969) indicates that root temperatures can affect root:shoot mass ratios, presumably through effects on physiological activity and hence carbohydrate partitioning (see Chapter 5). Soil temperatures also affect the activity of micro-organisms and hence the rate of breakdown of organic matter in the soil—provided that the soil is wet enough for biological activity. Soil freezing and thawing may be very important, particularly through effects on the time at which growth commences in spring in cold climates.

2.3.3 Air Humidity

The partial pressure of water vapour in air saturated at a particular temperature (T) is called the *saturation water vapour pressure* ($e_s(T)$) and depends on temperature alone. Saturation vapour pressure (kPa)[†] may be calculated from a number of equations; a simple and accurate expression is given by Dilley (1968):

$$e_s(T) = 0.61078 \exp[17.269T/(T+237.3)]. \quad (2.10)$$

The temperature at which water begins to condense out of a volume of saturated air is called the *dew point* (T_d).

The partial pressure of water vapour in unsaturated air (usually called simply *vapour pressure*, e_a) can be calculated from the standard psychrometric equation

$$e_a = e_s(T_w) - \gamma(T - T_w) \quad (2.11)$$

where T_w is the wet-bulb temperature, and $e_s(T_w)$ is saturation vapour pressure at T_w. For fully ventilated psychrometers the psychrometric constant (γ) has the numerical value 0.066 kPa°C^{-1}. Full ventilation implies that radiative exchanges between the sensors and the air stream are negligible and that forced convection is the dominant heat exchange mechanism between sensors and the air. Therefore the sensors measure true air temperature; the rate of evaporation from the wet-bulb element depends only on the vapour pressure deficit of the air and any increase in the flow rate does not cause an increase in the wet-bulb depression ($T - T_w$).

The ratio of the vapour pressure of unsaturated air (e_a) to saturated vapour pressure of air at the same temperature ($e_a/e_s(T)$) is the *relative humidity* (RH), usually expressed as a percentage. Relative humidity only

[†] The standard unit of pressure is the pascal (Pa). Vapour pressures are nowadays usually expressed in kPa. In a great deal of earlier literature the millibar was the standard unit used; the conversions are straightforward: 1 millibar = 100 Pa hence 1 kPa = 10 millibars.

provides information about the absolute value of humidity if the temperature is also specified. Some biological phenomena, such as the opening of seed cases, are affected by the RH. It also influences the likelihood of forest fires—and their severity if they occur—because dry plant material absorbs water if relative humidity is high. The amount of water in such material has a very significant effect on the ease with which it will ignite and the temperature with which it will burn.

The *humidity mixing ratio* (the mass of water vapour per unit mass of dry air with which the water vapour is associated) is

$$r = \frac{\varepsilon e_a}{P - e_a} \tag{2.12}$$

where ε is the ratio of the molecular masses of water vapour and dry air ($=0.622$) and $P=$ atmospheric pressure. Since e_a is small compared to P (if P is in kPa, standard atmospheric pressure $= 101.3$ kPa) then r will be of the order of 0.01 g g^{-1} and would more conveniently be expressed in grams water vapour per kilogram air. Approximate values of r for $P = 100$ kPa are given by

$$r \approx 6.2 e_a \quad \text{g kg}^{-1}. \tag{2.13}$$

The humidity mixing ratio has the advantage of being a conservative measure of water vapour concentration—it is independent of temperature and pressure. The *absolute humidity* of air (χ, kg m^{-3}) is obtained by multiplying r by air density ρ_a (≈ 1.2 kg m^{-3}). It is affected by both temperature and pressure and hence should be avoided.

Water moves across vapour pressure gradients, and when we consider transpiration, the vapour pressure difference between the inside of the leaves and the ambient air (δe) is the important variable. It depends on the vapour pressure of the air and on foliage temperature (T_f)—assuming that air in substomatal cavities is saturated with water vapour:

$$\delta e = e_s(T_f) - e_a. \tag{2.14}$$

The *vapour pressure deficit* (D) is an important measure of the drying power of the air:

$$D = e_s(T) - e_a. \tag{2.15}$$

Air humidity is determined by air mass characteristics and is affected by evaporation from underlying surfaces. Diurnal variations are difficult to predict but to obtain estimates of D and (if required) RH, we can assume that e_a is constant through the day, with the dew point temperature equal to T_{min} (which usually occurs in the early hours of the morning before dawn), i.e. $e_a = e_s(T_{min})$. It follows that, given measured or estimated temperature at any time (equation (2.9)), estimates of D and RH can be calculated.

2.3.4 Wind Speed

Wind is important to plants and plant communities because it is responsible for the turbulent transfer of heat, water vapour and carbon dioxide (as well as dust, pollen pollutants and other substances). The turbulent transfer process is discussed in more detail in Chapter 3. It will suffice to note here that eddies and gusts are characteristic of wind and that wind damage to trees—a matter of considerable concern to foresters—often occurs because the frequency of gusts corresponds to the oscillation cycle of trees. This depends not only on wind force and characteristics, but also on tree height and diameter, the bending properties of the wood and the extent of mutual protection by trees. Damage to trees, such as breaking or uprooting, may also, of course, be caused simply by the drag force generated by wind.

2.3.5 Precipitation

Rainfall or any other form of precipitation (e.g. snow, hail) is important to plants as a means of replenishing soil water and maintaining soil water content high enough to permit uptake by plant roots. The average annual rainfall figures so often used in the belief that they convey useful information are, for many regions of the world, not very useful. Even in relatively damp, cool climates the amount of rain in a year may easily deviate from the long-term average by 50% while in semi-arid regions variations by a factor of two or even three are common. In cool temperate regions the variation in rainfall amounts may be relatively unimportant, especially if precipitation is increased by winter snow, but in dry regions, where the rate of water use by vegetation is high, drought (long rain-free periods) may be very serious. (Regions where forestry is important, and which are subject to drought, include parts of the United States, Chile, Brazil, Australia and South Africa.)

It is not only the total amount of rainfall that is important but its distribution in time. This influences the effectiveness with which soil water is replenished or maintained by a unit quantity of rain. Where the probability of precipitation is approximately the same throughout the year, serious water stress is less likely than where rainfall is unevenly distributed, not only because the soil water is frequently replenished but also because evaporation is reduced during rain periods. Precipitation effectiveness has to be evaluated in relation to evaporation. As a general rule the reliability of precipitation decreases as average precipitation decreases.

In regions where precipitation is strongly seasonal the growing season of trees may be greatly reduced by water shortage; good examples are regions with Mediterranean-type climates with winter rain. Here the growing seasons may be restricted to autumn periods, after rain starts but before temperatures fall too low for growth, and to spring/early summer when there is adequate water stored in the soil and rising temperatures allow growth.

Some rain comes in high intensity storms at rates which exceed the capacity of the soil to accept it. Such rain is likely to be less effective than lower intensity rain since much of it may be lost in run-off. On the other hand, intermittent light rain which falls on dense forests with high canopy water-holding capacity is also ineffective in that very little of it will reach the ground to replenish soil water, and if the canopy dries before the next shower there will be virtually no throughfall. (Canopy interception is considered in some detail in Chapter 7.) Since rainfall frequency, distribution and type all contribute to its effectiveness, evaluation of the potential of an area for forest productivity must include evaluation of the probability of precipitation in different seasons and should also include assessment of rainfall type, and hence effectiveness. Effectiveness can only be properly evaluated by calculations using the hydrologic equations (equations (7.29) and (7.30)), which lead to estimates of the soil water balance. The relationships between soil and plant water status, and plant growth, are discussed in Chapter 7.

2.4 CONCLUDING REMARKS

This chapter has provided the basic framework for the rest of the book, where ideas about rates—which are greatly influenced by weather conditions—and states, underlie all discussions of processes. Forestry may be defined as the business of managing forests: it is concerned with forest communities, systems at the highest level of organization with response times of months to years. Management of both natural forests and plantations often involves actions—such as planting at a particular spacing, fertilizing, weed control or no control, and thinning—which disturb the system and cause a change in its state. Such actions are normally intended to achieve specified results, so the manager would like to be able to predict their consequences—the time course of subsequent change in the state of the forest. Changes in the state of forests affect the way they respond to weather conditions. Accurate prediction of the way forests respond to change or disturbance, and to weather conditions, will only be possible if the responses are understood in terms of the lower-level mechanisms and weather-affected processes involved. This implies that we have to understand, and be able to evaluate quantitatively, how processes such as carbon assimilation and allocation, nutrient uptake and cycling, and water relations affect the survival and growth of trees and hence changes in the state of forests.

The argument that predictions of the state of forests should be based on understanding weather-affected processes leads immediately to questions about the usefulness of such predictions, because of our inability to make accurate long-term weather forecasts. However, we can, in principle, combine our knowledge and understanding of the effects of weather on forests with stochastic models with which the probability of occurrence of various weather types and sequences can be calculated. We can evaluate the risks and probable upper and lower growth rates over particular periods.

The physical conditions actually prevailing at the sites of the physiological processes are determined by the interactions between weather conditions and the stand itself. These are treated in the next chapter.

3 Stand Structure and Microclimate

Weather affects every aspect of the growth cycle of trees and long-term observations of growth may be usefully analysed in terms of simple, unadjusted averages of weather variables, using values obtained from standard meteorological enclosures. However, conditions at the sites of the physiological processes that determine plant growth may be quite different from conditions in enclosures outside forests or even in large clearings or above canopies where measurements are usually made. The way the stand itself modifies weather conditions, to produce its own microclimate, must therefore be understood. This involves appreciation of the physical structure of forest stands and the way this structure is acted upon by, and interacts with, other physical factors such as radiant energy and air movement. Appreciation of the importance of stand structure in determining canopy microclimate leads immediately to (at least qualitative) appreciation of its importance in determining the responses of a forest to changes such as thinning—which alters element spacing and crown density—or fertilization (altering crown density).

In forestry terminology, stand structure is usually described in terms of tree height (perhaps including comment on the heights of dominant and co-dominant trees), canopy closure—usually merely to note whether it has occurred or not—and crown depth. More important, from the forester's point of view, are measures such as total standing basal area, tree number and stem diameter. On the basis of height, diameter and basal area, estimates of standing timber volume are made using empirical equations.

Although such measures of stand structure are useful to the operational forester they have limited value for analysing the growth of trees in terms of their response to environmental factors. For this purpose the most important characteristics of forest stands are their leaf area index and leaf area density; these determine the amount of radiation that will be absorbed by the stand and hence the amount of CO_2 taken up in photosynthesis. They also greatly influence the amount of water vapour transpired and the aerodynamic exchange processes in the canopy.

Stand structure is almost invariably described in static terms—the state of the stand at any given time—but canopy dynamics are an integral part of the growth patterns of forests. Changes in canopy structure depend, in the short term, on leaf dynamics, and, on longer time scales, on changes in stem numbers, which may involve competition between stems for water and nutrition. In an even-aged forest or plantation there will be a steady decline in stem numbers as trees die. In uneven-aged forests, particularly those with mixed species, there is constant replacement of the trees that die and the gaps left are exploited by young trees. In this chapter we will be concerned largely with even-aged forests, although in many respects the discussion is completely general.

Radiant energy is the "driving variable" for photosynthesis, and hence carbon assimilation, so the absorption and utilization of energy by forests is a fundamental determinant of forest productivity—it sets the upper limits to growth. Energy absorption depends on canopy structure and density and is treated in this chapter. Leaves utilize energy to convert CO_2 and water vapour to carbohydrate in the process of photosynthesis, i.e. they convert electromagnetic to chemical energy. Leaf photosynthetic characteristics are discussed in Chapter 4.

Radiant energy, air humidity and turbulent transfer determine the amount of water lost by leaves, and hence by trees. The coupling between the aerial environment inside canopies, and the trees, is largely through the leaves, and can be described by the leaf energy balance. Turbulent transfer processes and the leaf energy balance are formally treated in this chapter. Equations for calculating transpiration are also presented but discussion of water balances is postponed to Chapter 7, where the water relations of trees are considered.

3.1 STAND STRUCTURE

3.1.1 Foliage Mass, Area and Distribution

The leaf area index (L^*) of a stand (m² leaf per m² land surface) depends on leaf area per tree (A_f, m²) and the tree population (p, trees per m²); i.e.

$$L^* = \sum_0^p A_f. \tag{3.1}$$

In the early stages of growth, before full canopy has been reached, L^* is not a particularly useful concept since individual trees may have quite dense foliage while there are gaps between trees. The important parameter in this case is leaf area density (ρ_f, m² m^{-3}) within the individual trees. The simplest assumption about the distribution of leaves within the canopies of individual

3 Stand Structure and Microclimate

trees is that they are randomly distributed with a constant probability of occurrence throughout the volume of the canopy, and that leaf orientations and azimuths are also random. In fact these assumptions are unlikely to be true for many trees species, where the foliage may be clumped in various ways. They may lead to significant errors in detailed analyses of lower order processes (e.g. leaf photosynthesis or transpiration at specified levels in canopies) but are unlikely to lead to significant error when working at higher levels of organization over long periods (e.g. canopy photosynthesis over periods of days or weeks).

The time required for a community of young trees to reach full (closed) canopy depends on the initial population, the species and growing conditions.

We would expect the foliage mass carried by individual trees to be proportional to their size and since foresters have, for a long time, used stem diameter (d_B) as a measure of size it follows that it should be possible to estimate leaf mass from stem diameter. This is well supported by the allometric relationships calculated from many biomass studies. These are usually of the form (cf. equation (5.8)).

$$\ln W_f = \ln c_f + n \ln d_B$$

i.e.

$$W_f = c_f d_B^n \qquad (3.2)$$

where W_f is foliage mass, but might be stem, branch or even root mass (see Chapter 5). A simpler relationship emerged from a study by Grier and Waring (1974), in which they established a linear relationship between the foliage mass W_f of conifers and sapwood cross-sectional area (A_s):

$$W_f = aA_s - c \qquad (3.3)$$

where a and c are constants. Equations (3.2) and (3.3) allow estimates of A_f if the specific leaf area (σ_f, leaf area per unit leaf weight) and the values of the constants are known. (σ_f will vary with species but may also vary with age and growing conditions.) Later studies have provided values of a (equation (3.3)) for a range of both coniferous and deciduous species (see Gholz et al., 1979). No exceptions have been found to the relationship although Pook (1984b) found that a quadratic equation gave the best fit to *Eucalyptus maculata* data. There have been few studies on seasonal variation (see 3.1.2, "Leaf dynamics"), the response times and feedback relationships between foliage production and increase in A_s and the variation of the parameter values for a particular species with site and fertilization. Pook followed seasonal changes in leaf area of *E. maculata* and pointed out that "basal area as an estimate of canopy leaf area is obviously insensitive to seasonal changes in leaf area associated with canopy renewal (but) the leaf area of eucalyptus overstorey determined from basal area may be expected to provide a useful benchmark for seasonal change in leaf area estimated from the canopy leaf area balance (leaf production and loss)". The most complete evaluation of equation (3.3) is that of Whitehead (1978), who not only showed that the values of *a* and *c* for Scots pine were constant over a range of stand densities, but also analysed the vertical distribution of the foliage.

Kinerson *et al.* (1974) in their analysis of the foliage dynamics of Scots pine, fitted a normal curve to the vertical distribution of foliage area density with height (z). Whitehead (1978) found for Scots pine that differences between plots in foliage distribution with height were described by differences in the standard deviation (s_d) of this distribution; this parameter is inserted in the equation for the normal distribution:

$$\rho_f = \frac{\Delta z L^*}{s_d (2\pi)^{1/2}} \exp\left(\frac{-0.5(z-\bar{z})}{2s_d^2}\right). \tag{3.4}$$

$\rho_f(z)$ is the foliage area density in the height interval z spanning height z, the mean height of the foliage is $\bar{z}$. Whitehead's data showed s_d varying from 1.36 for a plot with 608 trees ha^{-1} to 1.05 for a plot with 3281 trees ha^{-1}. Although there have been only a limited number of studies of this sort, it is likely that (3.4) will often give a reasonable description of foliage density distribution. However, Norman (1979) has found that for purposes of radiation modelling foliage area distribution can generally be adequately described by two straight lines intersecting at the height (z_{max}) where ρ_f is greatest (see Fig. 3.1).

A series of leaf area profiles given by Rauner (1976) for deciduous forests supports the assumption that foliage density distribution is usually normal with height, although distributions that are biased significantly in both directions occur in different species.

Equations (3.2) or (3.3), and (3.4) provide a fairly complete description of a forest canopy. However, the information required for a particular species and different stand densities will usually have to be collected.

The leaf area index of the four populations studied by Whitehead (1978) was essentially the same (2.5–3). The basal (trunk) cross-sectional area A_b decreased by a factor of about 2 from the lowest to the highest population and the ratio A_s/A_b also decreased (by about 15%). The tree communities studied by Whitehead were about 40 years old and had probably reached

3 Stand Structure and Microclimate

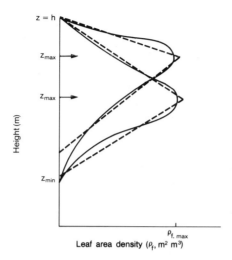

Fig. 3.1. Typical foliage area distribution curves for forests (see Kinerson et al., 1974; Rauner, 1976). The curves can be described by the equation for a normal distribution (equation (3.4)) but linear approximations (broken lines) are easier to handle and will generally be adequate for calculating energy absorption by canopies (Norman, 1979). Values of $\rho_{f\,max}$ may be up to 1.5–2 $m^2\,m^{-3}$. Integrating ρ_f from $z=h$ to z_{min} gives L^*; if the linear approximations are used then $L^*=1/2\rho_{f\,max}(h-z_{min})$.

some sort of equilibrium, although this would not be stable in the long term. Waring et al. (1981) used an experiment in which thinning treatments had been imposed on a natural stand of Douglas fir (*Pseudotsuga menziesii*), with stand density maintained by successive thinnings. The trees were about the same age and the same size as those studied by Whitehead but L^* varied from 3.6 to 12, with populations of 173, 304, 477, 593 and 1977 trees ha^{-1}. The ratios A_s/A_b did not vary significantly. These thinned stands were clearly maintained in an unstable state and had not reached equilibrium.

The values of L^* achieved by natural forests can vary enormously. Grier and Running (1977) suggest that L^* values (on a projected leaf area basis) of 18 to 20 appear to be about the upper limit for coniferous forests; the data for temperate deciduous forests which they reviewed suggested L^* of about 6 as a typical value. They established a remarkably consistent negative linear relationship between L^* and a seasonal water balance index calculated by adding soil water storage to measured growing season precipitation and subtracting open pan evaporation (see Chapter 7). The water balance index accounted for 99% of the variance in L^*, but the relationship has not been tested for trees in other areas.

Grier and Running do not say whether the changes in L^* are primarily caused by reductions in tree size or tree number. However, their results are consistent with the general observation that both tree populations and leaf area per tree decrease with increasing aridity of the environment. This point is also made by Anderson (1981), who tabulated published values of L^* for forests, the lowest of which was 2.6 (young *Fagus sylvatica* in Denmark) to compare with a series of determinations of L^* for eucalypts in south-east Australia. The average value for the eucalypts was 1.2. These observations suggest that the slope of the relationship between A_f and A_s (a in Equation (3.3)) is likely to be less for trees subject to water deficits than for those not short of water, i.e. less foliage per unit sapwood area. It follows that resistances to the flow of water through trees of similar size and sapwood cross-sectional area are likely to be larger in trees growing in dry areas (see Fig. 3.2 and Chapter 7). This is consistent with the suggestion by Jarvis

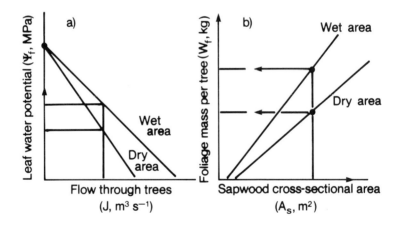

Fig. 3.2. Diagrammatic representation of the consequences of differences in resistance to flow through trees, and in the slope of the relationship between sapwood cross-sectional area and leaf mass. (a) If resistance to flow through the conducting tissues of trees in dry areas is higher than in wet areas, a given flow rate (J) will result in lower tissue water potentials in "dry area" trees than in "wet area" trees (see equation (7.13) and Fig. 7.3). (b) We would expect a given sapwood cross-sectional area (A_s) to support less foliage in a dry area than a wet area (see equation (3.3)). Note that these effects could cancel one another.

(1975) that there is homeostasis between sapwood area and leaf area operating so that, in a given environment, leaf water potential will seldom fall to damaging levels. Whitehead and Jarvis (1981) also point out that the resistance to flow through trees will be much larger in high-population stands than in low-population stands with the same growing conditions.

These considerations suggest that the "full canopy", regarded by foresters as the normal desirable situation for a community "fully exploiting the site" may in fact not be a normal situation for forests in areas subject to severe and consistent water deficits. Full canopy tends to be a qualitative description and it is unlikely to be attained until L^* has a value of about 5. If it is attained by high populations in the early stages of, say, stand regeneration, the population will be unstable and will shift towards lower stem numbers and L^* as the stand matures.

3.1.2 Leaf Dynamics

Equations (3.1) and (3.3) describe the relationships between stem number and leaf area of a forest stand at any time, i.e. the equations provide estimates of the state of the stand in terms of those variables. However, the leaf area of trees—and hence of forest stands—is seldom constant for long. The life of a leaf may vary from less than a year on deciduous trees in cool climates to many years on conifers in cool wet regions. Leaves on deciduous trees such as eucalypts may survive for two to three years.

In principle leaf area dynamics can be described by a differential equation. The net rate of change of leaf area is

$$dA_f/dt = \text{rate of increase} - \text{rate of loss.} \qquad (3.5)$$

Clearly on a tree increasing in size, the increase of leaf area over a relatively long interval, such as a year, is likely to exceed the loss since the class of new leaves produced on new shoots, and to replace those that have fallen from older shoots, will be larger than the class of old leaves falling. (This will only hold if the tree is not subjected to some stress such as unusual drought, which accelerates leaf fall. Insect attack may also, of course, be a major disruptive factor.) The expectation that leaf production is proportional to growth leads to the further expectation that leaf fall—commonly measured in litter fall and nutrient cycling studies—will be proportional to above-ground biomass increments in forests. This is confirmed by data collated by Miller (1984).

A tree or forest has reached steady state when the long-term rate of leaf production is approximately matched by rate of leaf loss, although it will not necessarily have a constant leaf area. This would only happen if both leaf growth and loss were more or less continuous. In fact a new generation of leaves is likely to be produced relatively quickly while a generation of leaves nearing the end of their life cycle may fall within a short period. Beadle et al. (1982) modelled annual changes in leaf area in Scots pine (in Scotland) using a set of linear regression equations based on litter fall measurements and leaf biomass determinations. Their analysis provides information on the general response times of the system studied, but no insights into the effects of changes in conditions on those response times. For this, more detailed studies are required. Beadle et al. (1982) found that all new leaves were produced between the beginning of June and the middle of August, that some of these (new) leaves were lost in the first year, losses continued at an increased rate in the second year and the remainder fell in their third year. Pook (1984a) provided a comprehensive account of the foliage dynamics of *Eucalyptus maculata*. He found that, in this species, foliage production may occur at any time of year, patterns of leaf production and leaf fall were variable—peaks were generally synchronized, with leaf fall lagging behind production.

Equations (3.6) provide a complete description of leaf dynamics. If we assume that the production of new leaves (class $A_{f.o}$) depends on tree size (given by stem-cross-sectional area, A_b), on the population of leaves already on the tree (age classes $A_{f.i}$) and on environmental factors (F_{env}) then, for the nth growing season, which covers the period from t_n-1 to t_n, $A_{f.i}$ is obtained from

$$dA_{f.o}/dt = f(A_{f.o}(t), A_{f.1}(t), A_b(t), F_{env})$$

$$dA_{f.1}/dt = f_1(A_{f.1}(t), A_b(t), F_{env}) \qquad (3.6)$$

$$dA_{f.2}/dt = f_2(A_{f.2}(t), A_b(t), F_{env}).$$

The dependence of $dA_{f.i}/dt$ on the leaves already present on the tree reflects the fact that leaves are not only the source of assimilate, but growing leaves are also a sink for carbohydrates and nutrients.

If equations (3.6) were linear, i.e. $dA_{f.i}/dt = k_i A_{f.i}$, or $= A_{f.i}/\tau_i$ then $A_{f.i}(t) = A_{f.i}(t_{n-1}) \exp[(t_n - t_{n-1})/\tau_i]$ and the time constant τ_i may be of order weeks to months. In practice, although a linear approximation may be very useful, it will only hold if conditions are reasonably constant or follow a standard pattern. The functional form of the relationship should be determined experimentally but we would expect the solutions to produce graphs such as those shown in Fig. 3.3. It is unlikely that analytical solutions to the model could be found or would be particularly useful.

Obviously, a good model of leaf dynamics would not only be valuable for predicting the leaf area of a forest at any time, but also for predicting the contribution of leaves to the litter which falls on the forest floor, and hence to estimates of nutrient cycling.

3 Stand Structure and Microclimate

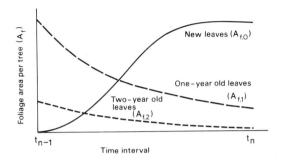

Fig. 3.3. Time course of change in leaf area in different age classes: the expected form of solutions to equations (3.6). See Pook (1984a) for observations of this type.

3.1.3 Stem Populations and Mass

From the point of view of light interception and canopy microclimate, foliage amount and distribution are the most important factors in stand structure but, from the point of view of commercial forestry, stem number and size (volume, mass) are the most important considerations. These also, of course, affect canopy microclimate directly, as well as in their interactions with foliage amounts, so although we will not be considering matters like stem size distribution, some consideration of how stem numbers may be affected by competition, and of the interrelationships between leaf area and stem populations, are pertinent here.

High populations of young trees will rapidly reach full canopy, although A_f may be small. There will then be progressive mortality ("self-thinning") with time, with the average size of individuals in the community tending towards a well defined maximum at any particular population (p), described by the empirical equation

$$W_{max} = k_* p^{-1.5}. \tag{3.7}$$

Equation (3.7) also describes the time course of self-thinning; when $W = W_{max}$ if W increases p must decrease or conversely W will not increase unless p decreases. The power -1.5 is extraordinarily stable across plant communities of all species (in pure stands) and ages (see White, 1981). The equation does not quantify plant mass if p is less than some limit value; W may then take any value less than that given by equation (3.7). White gives $k_* \approx 10^4$ g as a reasonable universal value for herbaceous plants where $p < 1$ m^{-2}.

When applied to trees, using volume units (m^3 tree^{-1}), with p in stems/hectare (10^4 m^2) k_* is 24 000 for radiata pine (Drew and Flewelling, 1977) and for *Eucalyptus regnans* and *E. obliqua* (P. West, personal communication). A rough analysis of *Pinus densiflora* data presented as a graph by Drew and Flewelling suggests that $k_* = 11\,000$–$15\,000$, and analysis of unpublished data of Lindsay (reported by Borough *et al.*, 1978) gives $k_* = 13\,000$. These values give $\ln k_*$ between 9.3 and 10.1, so the differences can probably not be taken as significant. If we convert from volume to mass units, using wood density values between 500 and 700 kg m^{-3}, so that W_{max} is in kilograms, we obtain a value of $\ln k_*$ of about 16.

Algebraic investigation of some of these relationships is interesting. We note that equation (3.2) relates foliage mass to sapwood cross-sectional area. Another well-established empirical relationship is

$$W = c_w d_B^{2.5} \tag{3.8}$$

where d_B is in centimetres. The power 2.5 relating tree mass (W) to stem diameter is entirely empirical. White (1981) tabulated data from a number of studies on many tree species which show this to be a very general value; there are, of course, quite large variations but no systematic departure from the value of 2.5 has so far been published. Combining equations (3.7) and (3.8), equating W and W_{max} and solving for d_B we obtain

$$d_B = k_*^{0.4} p^{-0.6} c_w^{-0.4} \tag{3.9}$$

so that, from (3.2), with $n = 2$

$$W_f = c_f \left(\frac{k_*^{0.4}}{p^{0.6} c_w^{0.4}} \right)^2 \tag{3.10}$$

and, converting W_f to A_f by specific leaf area (σ_f, m^2 kg^{-1}) we obtain the expression

$$A_f = c_f \, \sigma_f \frac{k_*^{0.8}}{p^{1.2} c_w^{0.8}} \tag{3.11}$$

or with

$$c_f \sigma_f k_*^{0.8} c_w^{-0.8} = C_f$$

where C_f is a composite constant

3 Stand Structure and Microclimate

$$A_f \approx C_f p^{-1.2} \tag{3.12}$$

and since (from (3.1))

$$L^* \approx A_f p$$

then

$$L^* \approx C_f p^{-0.2}. \tag{3.13}$$

We have therefore arrived at an expression for leaf area in terms of the population of trees, provided there is sufficient competition within the population to cause self-thinning. Since the value of the exponent of p depends on the power 2.5 in equation (3.8), it could deviate significantly from -0.2, but this is somewhat irrelevant.

Some numerical values for the parameters of equations (3.9) to (3.13) may be helpful. "Reasonable" values are: $c_f = 200$ kg, $c_w = 5700$ kg, $k_* = 8.9 \times 10^6$ kg and $\sigma_f = 3$ m^2 kg^{-1}. These give $C_f = 2 \times 10^5$ and, taking $p = 200$, 500 and 2000 stems ha^{-1}, $A_f = 346$, 115 and 21 m^2 (from equation (3.12)) and $L^* = 6.9$, 5.7 and 4.2 (from equation (3.13)).

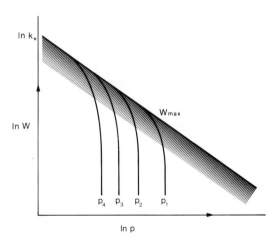

Fig. 3.4. Diagrammatic illustration of equation (3.7). W_{max} is the maximum average mass of trees in population i, attained by self-thinning. The W_{max} line has a slope of $-\frac{3}{2}$. The p_i curves indicate that the mass of trees in a population can increase, independent of population, until significant competition begins in the zone of imminent competition (shaded), after which W_i tends to $W_{max}(p_i)$. Thinning moves tree populations to lower p.

In view of the way L^* appears to vary with the water balance of sites (Grier and Running, 1977) we would expect that k_* (in equation (3.7)) would also vary. It may be possible, given sufficient data on W and p for mature forests, to demonstrate that k_* is a function of site water balance—we would expect it to be highest in regions normally having large positive water balance, and vice versa. This suggestion is supported by the fact that assimilate production is proportional to radiant energy absorbed and the result of low L^* will be small amounts of energy absorbed.

It is worth noting, in passing, that equation (3.7) has considerable implications for thinning practice. If, for a population p_i, $W_i < W_{max}$, the growth rate of the trees (dW/dt) will be essentially independent of population until W moves into the area of imminent competition (Fig. 3.4), where dW/dt under particular environmental conditions becomes influenced by competition. If a stand is thinned before this the period before competition becomes a significant factor will be extended. The value of W_{max} depends on site (fertility, soil water holding capacity) and climate, particularly the water balance. It follows that when self-thinning occurs the growth rate of individual trees in a stand will be reduced.

3.2 STAND MICROCLIMATE: ENERGY RELATIONSHIPS

The gross features of stand structure in a mature forest that has been allowed to reach equilibrium with its environment are strongly dependent on climate. On shorter time scales the physical characteristics of a stand interact with weather to determine the physical environment (microclimate) inside the stand. This in turn exerts a major influence on physiological processes and hence growth patterns.

Treatments of the principles of environmental physics and microclimatology have been presented in recent years in books by Monteith (1973) and Campbell (1977). The micrometeorology of coniferous forests has been comprehensively reviewed by Jarvis et al. (1976) and some aspects of the micrometeorology of deciduous forests by Rauner (1976). In this section, therefore, only the important features of forest microclimate will be outlined. Unless otherwise stated the discussion refers to forests with closed canopies.

To conform to the pattern of earlier chapters, we should begin the treatment of stand microclimate by considering the absorption and partitioning of energy by the basic elements of a stand—the leaves. Individual leaves may respond to changes in energy load in seconds (see Fig. 2.6); small changes in air temperature in canopies may occur across similar periods while significant changes associated with the interruption of clear sky conditions by the passage of clouds, for example, may occur in minutes. These changes are superimposed on diurnal cycles in air and tissue temperatures. Leaves are also the main organs responsible for absorption of momentum, hence the distribution of leaf area density determines the shape of wind profiles in canopies. However, before discussing leaf energy balance, we will consider canopy energy balance and partitioning and then radiation penetration into, and absorption by, canopies.

3.2.1 Canopy Energy Balance

The energy balance of a forest canopy (or any surface) results from radiative and dissipative processes, that is from the relative rates of input and loss of radiant energy, and the rate of dissipation of absorbed energy. The net radiation φ_n retained below any surface is

$$\varphi_n = (1 - \alpha)\varphi_s + \varphi_L \tag{3.14}$$

where α, the albedo of the surface, is the fraction of incident short-wave radiation which is reflected (α is also called the reflection coefficient, or reflectivity, of the surface) and φ_L is the long-wave balance.

Any body with a temperature greater than absolute zero loses energy by radiation at a rate proportional to fourth power of its absolute temperature, i.e. $\varphi_L = \varepsilon_L \sigma T^4$ where σ is the Stefan–Boltzmann constant ($= 5.67 \times 10^{-8}\,\mathrm{W\,m^{-2}\,K^{-4}}$), T is the absolute temperature and ε_L is the emissivity. For a perfect black body $\varepsilon_L = 1$. For most vegetative surfaces ε_L is in the range 0.90–0.96. The downward flux of long-wave radiation depends on the apparent radiative temperature of the sky; upward flux depends on the effective temperature of the radiating surface. (Upward fluxes are conventionally denoted as negative, downward as positive.)

As a general rule, the reflectivity of surfaces decreases with their roughness, and since forest canopies are rough surfaces their reflectivity is low. Figure 3.5 is derived from Stanhill (1970) and illustrates changes in albedo with the height of the elements of natural surfaces. Representative (mean daily) values of α are about 0.10 for coniferous forests (Jarvis et al., 1976) and 0.16 for deciduous forests (Rauner, 1976). There have been several thorough studies of the radiation balance of forests, e.g. those by Moore (1976) on *Pinus radiata* in South Australia, and Stewart and Thom (1973) on *P. sylvestris* in England. Moore identified variations in albedo through the day but these were relatively trivial and he concluded that a mean albedo of 0.11 ± 0.01 could be applied to the canopy for all solar elevations. Pinker (1982) found that the albedo of a tropical evergreen forest showed very

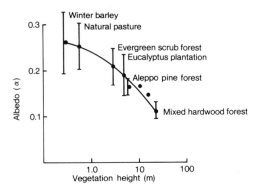

Fig. 3.5. Relationship between mean albedo and the height of vegetation. The vertical lines correspond to two standard deviations (redrawn from Stanhill, 1970). (Note that the *x* axis (vegetation height) is a logarithmic scale.)

marked diurnal variation, from about 0.17 (early morning/evening) to 0.11 midday. There were also seasonal differences.

For many purposes φ_n can be conveniently estimated from empirical relationships of the form

$$\varphi_n = a_n + b_n \varphi_s. \quad (3.15)$$

This can be established from measurements of φ_n over a forest and φ_s need not be measured nearby. Jarvis *et al.* give values of b_n for coniferous forests which range from 0.71 to 0.91; a value of 0.8 will seldom be much in error although Moore's study is again worth consideration. He gave $\varphi_n = 0.67\,\varphi_s - (45 \pm 10)$ W m^{-2} for the winter months and $\varphi_n = 0.85\,\varphi_s - (55 \pm 18)$ W m^{-2} for summer. Federer (1968) found $b_n = 0.83$ for a hardwood forest while Rauner (1976) indicated that 85% of φ_s was absorbed by a dense aspen stand (i.e. $b_n \approx 0.85$) and Kalma and Fuchs (1976) give $b_n = 0.8$ for a citrus orchard.

a_n gives an estimate of the average value of φ_L and thus depends on local climate as well as forest structure. The value of a_n for coniferous forests varies widely, from -6 to -126 W m^{-2}; the median value is about -60 W m^{-2}. There are few values available for deciduous forests other than that given by Federer (1968): $a_n = -89$ W m^{-2}.

3.2.2 Partitioning Absorbed Energy

Conservation of energy requires that φ_n be partitioned into latent heat

3 Stand Structure and Microclimate

(evaporation or transpiration λE_t), sensible heat (H), absorbed by the adjacent air, and heat stored below the surface where φ_n is measured (G):

$$\varphi_n - \lambda E_t - H + G = 0. \tag{3.16}$$

Over any 24 h period G will usually be negligible so, although it should not be ignored in detailed short-term analyses, we will for simplicity neglect it in the following analysis.

Energy partitioning is often described by the ratio of H to λE_t, known as the Bowen ratio:

$$\beta = H/\lambda E_t. \tag{3.17}$$

Substituting this in (3.16) (ignoring G) we obtain

$$\lambda E_t = \varphi_n/(1+\beta) \tag{3.18}$$

and

$$H = \beta \varphi_n/(1+\beta). \tag{3.19}$$

Therefore, given values for φ_n and β, we can obtain estimates of λE_t.

The principal advantage of the Bowen ratio is that it is relatively straightforward to determine experimentally, although the measurements required are technically demanding. The basis of the method lies in the fact that we may write λE_t and H as the fluxes of water vapour and heat in terms of gradients of water vapour pressure and temperature. The flux of an entity is proportional to the product of its concentration gradient ($d[c]/dz$) and a turbulent exchange coefficient K:

$$\text{Flux} = K\, d[c]/dz.$$

The air flow regime above forests is invariably turbulent, so K is typically several orders of magnitude larger than molecular diffusion coefficients. λE_t and H are given by equations (3.20) and (3.21):

$$\lambda E_t = \frac{\rho_a c_p}{\gamma}(de/dz) K_V \tag{3.20}$$

$$H = \rho_a c_p (dT/dz) K_H \tag{3.21}$$

where γ is the psychrometric constant, K_V and K_H are the appropriate turbulent exchange coefficients and ρ_a and c_p are air density and specific heat respectively. Now substituting (3.20) and (3.21) into (3.17), and assuming $K_H = K_V$ we obtain

$$\beta = \gamma \, \Delta T / \Delta e \qquad (3.22)$$

where ΔT and Δe are measured across the same height interval Δz. β can therefore be determined by estimating the gradients of temperature and vapour pressure from measurements above a canopy.

Jarvis et al. (1976) summarized the values of β determined over coniferous forests and found that for most forest sites, irrespective of species, β for dry canopies varies between 0.1 and 1.5 (implying $\lambda E_t/\varphi_n$ varies from about 0.9 to 0.4) and, when the canopy is wet with rain or dew, between -0.7 and $+0.4$ (negative values of β indicate that energy is being supplied to the surface). However, β may rise to values of about 5 or even 10 for pine forests during the day (Stewart and Thom, 1973; see Fig. 3.6).

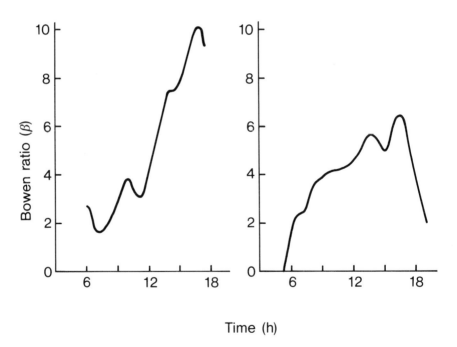

Fig. 3.6. Diurnal course of variation in the Bowen ratio (β) over a Scots pine forest in Norfolk, England, during two consecutive days (2–3 June 1971). Both days were fine and clear, with available energy ($\approx \varphi_n$) reaching 600 W m^{-2} at midday (redrawn from Stewart and Thom, 1973).

3 Stand Structure and Microclimate

Since β is the ratio of sensible to latent heat fluxes and high values over vegetation occur during the day, they must be associated with periods when latent heat flux (transpiration) is low, which is likely to be a consequence of stomatal control (see Chapter 4).

There seem to be few, if any, direct measures of β for deciduous forests, but Rauner (1976) gives average values of φ_n and λE_t for a number of such forests and since equation (3.18) may be rewritten

$$\beta = \varphi_n/\lambda E_t - 1$$

β can be calculated from these. Rauner gives integrated data for summer for deciduous forest types in northern Europe from which we obtain the average values of $\lambda E_t/\varphi_n$, and hence β, given in Table 3.1. (Note that, because β varies in non-linear fashion during the day, average values should be treated with considerable caution.)

Table 3.1. Average summer values of $\lambda E_t/\varphi_n$ and β for several deciduous forest types in Europe (from Rauner, 1976).

	$\lambda E_t/\varphi_n$	β
Mixed deciduous forest (*Betula verrucoso, Populus tremula*)	0.94	0.06
Oak (*Quercus robur*)	0.84	0.19
Maple plantation (Acer spp.)	0.88	0.14
Linden, oak (*Tilia cordata, Quercus robur*)	0.7	0.37

3.2.3 Radiation Penetration into and Absorption by Canopies

The mathematical description of short-wave radiation penetration into plant stands can be complicated; rigorous treatment involves knowledge of leaf angle distribution, leaf clumping characteristics, sun angle and the proportions of solar radiation which are direct beam or diffuse. Ross (1976) provides a good treatment and Norman (1982), has provided a simplified version of his own thorough earlier treatments. This is adequate for virtually all calculations of photosynthesis, which require the flux densities of photosynthetically active radiation.

The following equations come from Norman (1982), who divides the canopy into two classes of leaves—sunlit and shaded. Direct beam ($\varphi_{s.B}$) and diffuse ($\varphi_{s.D}$) radiation are intercepted very differently by canopies, so they are treated separately, which implies that they should be measured separately or the ratio $\varphi_{s.B}/\varphi_s$ (where $\varphi_s = \varphi_{s.B} + \varphi_{s.D}$) must be documented for different seasons and for particular locations. For canopies with spherical leaf angle distribution which implies that, from whatever angle radiation comes, there is always an equal leaf area normal to the beam, sunlit leaf area index is given by

$$L^*_{sun} = [1 - \exp(-0.5L^*/\cos Z)]2\cos Z \qquad (3.23)$$

and the shaded leaf area index by

$$L^*_{shade} = L^* - L^*_{sun}. \qquad (3.24)$$

Z is the solar zenith angle, which depends on latitude, date, time and solar declination. It can be calculated or obtained from charts (see List, 1968). The average visible irradiation received by all shaded leaves ($\varphi_{f.shade}$) can be approximated by

$$\varphi_{f.shade} = \varphi_{vis.D} \exp(-0.5L^{*0.7}) + S_c \qquad (3.25)$$

$\varphi_{vis.D}$ denotes the diffuse component of the visible radiation incident on the canopy ($\varphi_{vis} \approx 0.5\varphi_s$). S_c arises from multiple scattering of direct beam (visible) radiation ($\varphi_{vis.B}$) and is given by

$$S_c = 0.077\varphi_{vis.B}(1 - 0.09L^*)\exp(-\cos Z).$$

The radiation received by sunlit leaves is given by

$$\varphi_{f.sun} = \varphi_{vis.B} \cos \alpha_i/\cos Z + \varphi_{f.shade} \qquad (3.26)$$

where α_i is the mean leaf–sun angle (60°, independent of sun-angle for canopies with spherical leaf angle distribution).

More simply—and subject to greater error—the transmittance (ζ) of a canopy for the diffuse (φ_{diff}) and direct beam (φ_B) components of total short-wave radiation can be described by a simple exponential relationship:

$$\zeta = \exp(-k_\varphi L^*) \qquad (3.27)$$

where the extinction coefficient (k_φ) depends on species and canopy density values, and may range from 0.3 to 1.5. Data collated by Jarvis and Leverenz (1983) indicate that the average value of k_φ for both coniferous and deciduous forests is about 0.5. The average irradiance $\varphi_s(z)$ at any level z in a

3 Stand Structure and Microclimate

canopy may therefore be related to total irradiance above the canopy ($\varphi_s(0)$) by the expression

$$\varphi_s(z) = \varphi_s(0) \exp(-k_\varphi L^*(z)). \qquad (3.28)$$

Values for k_φ may be empirically derived by measuring $\varphi_s(z)$ at a number of levels in canopies and plotting $\ln[\varphi_s(z)\varphi_s(0)]$ against $L^*(z)$. If equation (3.28) holds this yields a straight line with slope of $-k_\varphi$. These measurements in forests pose tremendous sampling problems.

Equation (3.28) gives an estimate of *average* irradiance, at any level in the canopy, but in fact the intensity of radiation in canopies may vary enormously because of sunflecks, especially where there is a considerable component of direct beam radiation. In these circumstances, intensity distributions under moderate L^* tend to be bimodal with low-level background diffuse radiation and bright sunflecks. This distribution may be of considerable importance for leaf processes—such as photosynthesis—which depend non-linearly on radiation. Over short periods of time (e.g. 1 h) the errors involved in using the average irradiance in a layer to calculate photosynthesis may be considerable, and any productivity model which purports to provide good estimates of CO_2 uptake by canopies over such periods must use equations (3.23)–(3.26) or more detailed light interception models. However, if we are concerned only with daily totals then the use of Beer's Law (equation 3.28) to calculate irradiance in different layers from hourly values of $\varphi_s(0)$—to calculate average rates of CO_2 uptake—probably does not cause great errors.

To calculate accurately radiation interception by trees in non-continuous canopies would involve a complex model with provision for tree geometry (height, leaf area density, shape and spacing) and the time course of solar elevation, as well as information about the relevant amounts of direct and diffuse radiation. An analysis by Charles-Edwards and Thorpe (1976) illustrates many of the points involved. However, a useful general approximation can be obtained by partitioning the total short-wave radiation into the proportion which misses all the trees (transmitted through gaps, ζ_g) and the proportion ζ_f which strikes the trees and is transmitted through them. The amount of short-wave radiation reaching the ground (φ_{sg}) is then given by

$$\varphi_{sg} = \varphi_s(\zeta_g + \zeta_f) = \varphi_s(\zeta_g + \exp(-k'_\varphi L^*_i)) \qquad (3.29)$$

where k'_φ is the extinction coefficient within the tree canopy and L^*_i is the leaf area index of the individual trees, on the basis of the plan area of the individual crowns—not averaged over the whole ground surface area. A value of about 0.6 is given for k'_φ for apple trees by Jackson and Palmer

(1979), who suggested the above approach for orchards. The principles apply equally to spaced forest trees, and many broad-leaved forest trees would no doubt take k'_φ of approximately the same value.

Net radiation ($\varphi_n(z)$) at any level (z) in a stand depends on the penetration and reflection of short-wave radiation and the upward and downward fluxes of long-wave radiation from sky, leaves and soil. Equation (3.28) provides a reasonable approximation to the absorption of net radiation in the upper part of forest canopies but tends to break down near the ground because of the upward flux of long-wave radiation from the soil. The magnitude of this flux depends on soil temperature, which in turn is strongly influenced by soil wetness.

3.2.4 Leaf Energy Balance

The energy balance of leaves and exchange of energy between leaves and air is a major determinant of canopy microclimate. Ignoring heat storage, the radiant energy absorbed by leaves is partitioned into latent heat—increasing the humidity of the surrounding air—or sensible heat, increasing its temperature. The exchange processes between leaves and air depend on wind speed, leaf shape, arrangement (mutual shelter) and stomatal aperture. Introducing the subscript f to denote foliage, the energy balance equation for a leaf may be written

$$\varphi_{nf} = \lambda E_{t,f} + H_f \qquad (3.30)$$

which may be written in terms of the fluxes of water vapour and heat:

$$\varphi_{nf} = \rho_a c_p \left(\frac{(e_s(T_f) - e_a)}{\gamma(r_s + r_{bV})} + \frac{(T_f - T_a)}{r_{bH}} \right). \qquad (3.31)$$

The only symbols not defined earlier are r_s, r_{bV}, r_{bH}. These denote stomatal resistance and the boundary layer resistances for water vapour and heat respectively. The term resistance derives from the Ohm's law analogue, or Fick's law of diffusion, used as the basis for analysing the movement of entities such as water vapour and CO_2 out of and into leaves:

$$\text{Ohm's law} = \text{Current (amps)} = \frac{\text{Potential difference (volts)}}{\text{Resistance (Ohms)}}$$

or

3 Stand Structure and Microclimate

$$\text{Flux of diffusing gas} = \frac{\text{concentration difference}}{\text{diffusion resistance}}.$$

Resistances measure the effects of turbulence, leaf boundary layer thickness and stomatal aperture on photosynthesis and transpiration. The reciprocals of resistances ($1/r$ = conductance, g) are used for many purposes, although resistances are often more convenient for algebraic manipulation. Jones (1983b) has provided a much more comprehensive description than that given here of the theoretical basis, units and usage of these terms in transport equations.

Equation (3.31) illustrates the importance of stomatal resistance in the leaf energy balance. It can be solved for T_f or r_s—given values of e_a, T_a, φ_{nf} and r_{bH}—by iterative methods (see Landsberg and Butler, 1980; Grace, 1983). In the laboratory stomatal resistance is often evaluated from water vapour flux measurements while modern porometers have made it relatively easy to measure in the field. Values of r_s are likely to be higher by a factor of about 10 than those of r_{bV} and r_{bH}; r_s may range from 100 s m^{-1} upwards while r_{bV} and r_{bH} fall within a much narrower range—typically from 30–40 s m^{-1} for large leaves to about 10 s m^{-1} for small ones. Stomatal resistance is therefore the major factor influencing the partitioning of φ_{nf} into sensible and latent heat fluxes. Changes in r_s, other factors remaining constant, will cause changes in leaf temperature (T_f) and hence in $e_s(T_f)$, the leaf–air vapour pressure gradient and transpiration rates. The effects of r_s on photosynthesis are discussed in Chapter 4.

Stomata are under physiological control and are affected by light, air humidity (in most plants), CO_2 and leaf water status. The stomata in leaves at any level in the canopy will be affected by conditions at that level. There is some discussion in Chapter 4 of the factors influencing stomatal conductance; more complete information can be obtained from the literature (e.g. Jarvis and Mansfield, 1981). Information on the stomatal responses of tree species needs to be documented, particularly in terms of the effects of light, air humidity and leaf water status and, where possible, described mathematically (see, e.g., Jarvis 1976; Thorpe et al., 1980). Stomata may respond to changes in ambient conditions in minutes but values of r_s averaged over an hour may be used in leaf energy balance analyses, and models describing stomatal responses can be applied for similar periods.

Transfer of an entity across leaf boundary layers is by molecular diffusion. The thickness of the boundary layer depends on leaf size and shape, the wind speed and air turbulence. The relationship between r_b and wind speed (u) is almost invariably of the form shown in Fig. 3.7. Outside the laminar layer the transfer of entities (CO_2, heat, water vapour) is by turbulent diffusion (forced convection), a process orders of magnitude faster than molecular diffusion.

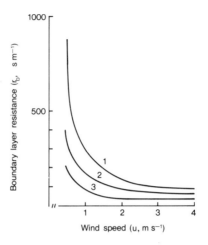

Fig. 3.7. Relationships between leaf boundary layer resistance and wind speed. The three curves are for leaves of different size: curve 1 is for leaves with characteristic linear dimension (CLD)=1 m, curve 2 for leaves with CLD=0.2 m and curve 3 for leaves with CLD=0.05 m, showing that r_b is greater for large leaves. For reference, wind speeds inside canopies are often much lower than 1 m s^{-1}, and the leaves of most forest trees have CLDs less than 0.05 m (redrawn from Grace, 1983).

Leaves with thin boundary layers (low r_b) are closely coupled to ambient conditions but very large leaves tend to have large boundary layer resistance, which may be a dominant factor in the leaf energy balance (see, e.g., Grace, 1977, 1981). Some of the implications of changing leaf temperatures and resistance values for transpiration rates are illustrated in Fig. 3.8.

3.3 STAND MICROCLIMATE: TRANSFER PROCESSES

3.3.1 Air Temperatures and Humidity in Forests

The air temperature inside a forest canopy may differ from that above the forest by several degrees and there may be significant gradients through the stand (see Fig. 3.9). Radiant energy intercepted by the canopy is partitioned into sensible and latent heat at plant surfaces, according to equation (3.31). The shape of the temperature and humidity profiles within the canopy depends on the distribution of foliage and its stomatal conductance and the

3 Stand Structure and Microclimate

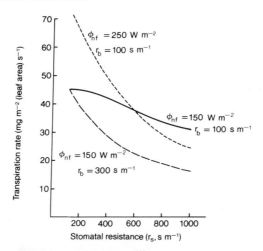

Fig. 3.8. Leaf transpiration rates, calculated for the φ_{nf} and r_b values given on the curves and a range of r_s values. Ambient conditions were specified as $T_a = 20°C$, $e_a = 1.5$ kPa; equation (3.31) was solved (by iteration) for T_f and hence $e_s(T_f)$. r_b denotes both r_{bv} and r_{bh}—assumed equal. At $\varphi_{nf} = 250$ W m^{-2} and $r_b = 100$ s m^{-1}, T_f increased from 29°C to 38°C. These increases in T_f, resulting in large increases in $e_s(T_f)$ and hence the leaf–air vapour pressure gradient (δe), kept transpiration rates from dropping as sharply as the increases in resistance alone would dictate.

level of turbulent transfer between the canopy and the air above. If there is a large area of foliage (high leaf area density, ρ_f) in any height interval Δz, absorbing large amounts of energy (φ_{nf} large), then the fluxes of sensible and latent heat into the "bulk" air in the canopy in that layer will be large. If ρ_f is high in the upper parts of a stand, then most of the momentum flux of the wind (see §3.3.2), will be absorbed in those layers, there will be little mixing of the air in the lower layers of the canopy with the overlying air and marked temperature and humidity gradients are likely to develop in the stand.

In a dense canopy with high leaf area density in the upper part, most of the incident energy will be absorbed in those layers, which will also absorb most of the momentum from the wind, causing low wind speeds and ineffective heat exchange throughout this canopy (see next section). The result will be higher temperatures in the upper rather than in the lower part of the canopy and high humidity if the stomata are open. Temperature and humidity in the lower part will depend very much on whether the soil is wet or dry. This applies particularly in open canopies (ζ_g large). Dry soil results in high soil-surface temperatures, hence high sensible-heat flux, but most of the energy

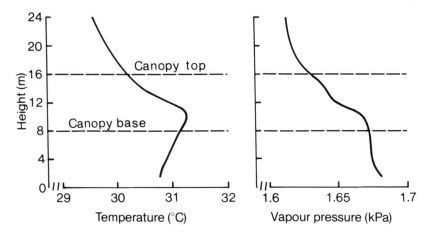

Fig. 3.9. Hourly average air temperature and vapour pressure profiles in a *Pinus ponderosa* forest in Australia. The highest temperature occurs in the region of highest leaf area density and air humidity also increases rapidly into that region (redrawn from Denmead, 1984). Profiles in other forest types, e.g. wet tropical or cold area coniferous forests, are likely to be very different. They will be influenced by foliage distribution and radiant energy absorption and loss, ambient temperature, humidity and wind conditions, and soil wetness.

falling on wet soil is converted to latent heat. The soil may then be cooler than the air and a sink for sensible heat, causing (in relatively dense canopies) higher humidity near the ground than in the upper part of the canopy (see right-hand profile in Fig. 3.9).

The albedo of the soil surface is also important. Albedo of bare soil is low (usually <0.10) while, if the canopy is sufficiently open to allow grass to grow, α may reach about 0.25. Reflection of this proportion of incident energy has important effects on the foliage energy balance.

It is apparent, therefore, that the temperature profile in open canopies must be strongly affected by the state of the soil and the heat flux from the ground. This also depends on wind speed and turbulence near the ground (see equations (3.16)–(3.21)). At night, when φ_n is negative, heat loss may be primarily from the upper part of dense canopies, or from the soil in open canopies.

Rigorous analysis of the processes of canopy microclimates is difficult because of variability, and our poor understanding of turbulent exchange

3 Stand Structure and Microclimate

there. The important point, which deserves emphasis, is the close coupling and feedback between the foliage and energy flows in the canopy.

Some of the transfer processes are discussed in more detail in the next section.

3.3.2 Turbulent Transfer Processes above Forests

Transfer processes between leaves and the air surrounding them have been discussed in terms of equation (3.31). The wind speed in the vicinity of leaves in a canopy, which governs their boundary layer (aerodynamic) resistance, depends on the air flow through the canopy as a whole. It is necessary to consider the factors affecting this flow, and the processes by which the properties of the bulk air inside the canopy are exchanged with the overlying air mass.

Much study of turbulent transfer processes above plant canopies, including forests, has been based on the assumption that these processes can be satisfactorily treated in one dimension (vertical). However, we must recognize that this is an approximation and that as soon as we move away from situations where it is (apparently) reasonable to assume horizontal homogeneity, we are faced with a chaotic and currently completely unquantifiable situation.

Turbulent diffusion can be envisaged as consisting of eddies—discrete parcels or bubbles of air, each of which has certain properties: momentum (mass × velocity), temperature (heat content), humidity (water vapour content), and CO_2 content. In air flowing over a rough surface, eddy size will tend to increase with distance away from the surface; near the surface the air is slowed by momentum absorption by roughness elements.

Eddies displaced either upwards or downwards will conserve their properties until they disintegrate and merge with the airstream at the level reached. Momentum transfer by eddies and momentum absorption by surfaces, and the consequent reduction of wind speed, cause wind profiles characteristic of the surfaces. This whole process of mixing and transfer of the properties of the air across concentration gradients, associated with turbulence, is of fundamental importance for plants and plant growth.

If the rate of air temperature decrease with height is near the adiabatic lapse rate ($0.01°C\,m^{-1}$), then when air flows over any reasonably level, uniform surface the vertical profile of horizontal wind speed will be logarithmic. (The wind profile shape depends on the vertical transfer of horizontal momentum, so it can be used to define an exchange coefficient K_M for momentum. If the rate of temperature decrease with height above the surface is appreciably greater than $0.01°C\,m^{-1}$, buoyancy effects will increase turbulence and alter profile shape.) The logarithmic profile is described by the equation

$$u(z) = \frac{u_*}{k} \ln\left(\frac{z}{z_o}\right) \qquad (3.32)$$

where u_* is the friction velocity, k is known as von Karman's constant and has the (empirically established) value of 0.41, and z_o is characteristic of the surface and relates to its roughness. The derivative of equation (3.32) is

$$du/dz = u_*/kz. \qquad (3.33)$$

For a given wind speed $u(z)$, u_* will be larger over a rough surface than a smooth one. Where a plant community (height h) causes the upward displacement of the wind profile a parameter d, the zero-plane displacement, has to be introduced and equations (3.32) and (3.33) become

$$u(z) = \frac{u_*}{k} \ln\left(\frac{z-d}{z_o}\right) \qquad (3.34)$$

and

$$\frac{du}{dz} = \frac{u_*}{k(z-d)}. \qquad (3.35)$$

Given values of d and z_o, and measurements of u at some suitable reference height, equations (3.32) and (3.34) provide a measure of u_* from which the momentum flux or drag force of the wind on the underlying surface, the shearing stress τ (kg m^{-1} s^{-2}), can be obtained

$$\tau = \rho_a u_*^2. \qquad (3.36)$$

The momentum exchange coefficient (K_M) is defined by the ratio of the momentum flux to its concentration gradient

$$K_M = \frac{\tau(z)}{\rho_a \, du/dz} = kzu_* \qquad (3.37)$$

or

$$K_M = k(z-d)u_*.$$

In the derivation of the Bowen ratio (equation (3.17)) we assumed that the exchange coefficients for heat and water vapour (K_H, K_V) were the same. For many purposes this is acceptable, but even if $K_H = K_V$ they cannot be taken as equal to K_M because thermal stratification of the atmosphere affects the exchange coefficients to different degrees.

The parameters z_0 and d characterize the aerodynamic properties of surfaces. They are normally derived from wind profile measurements made over large, homogeneous flat areas and as a first-order approximation may be expressed as fractions of the height of the roughness elements: $z_0 \approx 0.1\,h$ and $d \approx 0.7h$.

It seems intuitively likely that d will depend on community element density; in the limit where there are no large roughness elements (trees) in an area the zero plane displacement would depend only on the nature of the surface—say, grass. The introduction of widely spaced trees would disrupt profile development so that the theory outlined here would not hold, at least on the micro-scale, i.e. on a height scale of tens of metres. As the roughness elements become more closely spaced and dense, wind profiles develop above the canopy so formed until, in the limit, where element density is such that the surface is impenetrable to wind, d must equal h. However, for vegetated surfaces this is never approached and virtually all the experimental evidence suggests that profiles can be adequately described by equation (3.34) with d held constant and z_0 allowed to vary with the density of the roughness elements.

There are many values of these parameters published for trees. Jarvis *et al.* (1976) collated data for conifers, where d/h ranged from 0.61 to 0.92 and z_0/h from 0.02 to 0.26—although most of the values were less than 0.1. Leonard and Federer (1973) collated values for a number of species, mainly conifers but including hardwoods; d/h varied from 0.58 to 0.89 and z_0/h from 0.02 to 0.18. Leuning and Attiwill (1978) found d/h to be ≈ 0.79 and $z_0/h \approx 0.07$ for a eucalypt forest and Tajchman (1972) found z_0 for a forest of Norway spruce to vary with wind speed (see Fig. 3.10)—a phenomenon usually only observed with far more pliable canopies such as cereal crops. Tajchman (1981), discussing the variation in published values of z_0 and d for forests, emphasized the fact that both parameters may assume a wide range of values for the same height, and that they seem to be dependent on the structural characteristics of forests.

Garratt (1977) collated data from many sources to provide a valuable diagram (Fig. 3.11) showing the variation in z_0 with roughness element density; element density is defined as the element silhouette area normal to the wind per unit surface area occupied by each element. Using this definition, the shape and slenderness of the elements are accounted for.

The earlier discussions of canopy and leaf energy balances show that the partitioning of absorbed energy depends on surface properties. The transfer

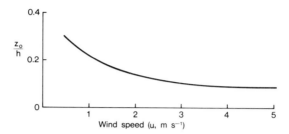

Fig. 3.10. Variation in the ratio z_o/h with wind speed above a Norway spruce (*Picea abies*) forest (calculated from data of Tajchman, 1972).

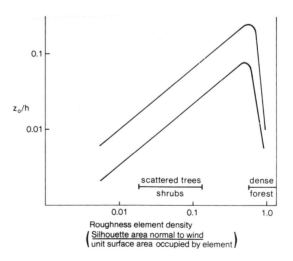

Fig. 3.11. Variation in the ratio z_o/h with roughness element density of natural surfaces. For scattered trees and shrubs roughness element density lies between 0.02 and 0.2; for a dense pine forest between 0.8 and 2.3 (redrawn from Garratt, 1977). (See Garratt's paper for methods of calculation and discussion.)

3 Stand Structure and Microclimate

of water vapour and heat depend on wind speed and the aerodynamic properties of the surface. Despite the possible differences in exchange coefficients, caused by atmospheric stability, a canopy resistance (r_{aM}) — analogous to r_b in equation (3.31) — can be calculated from the momentum exchange properties of forests and may be used in equations to estimate transpiration rates from canopies (see §3.4). This term derives from (3.36) which may be written

$$\tau = \rho_a u_*^2 = \rho_a u(z)/r_{aM}(z) \qquad (3.38)$$

Equation (3.38) may be rearranged to give the canopy conductance (g_{aM})

$$g_{aM} = 1/r_{aM} = u_*^2/u(z). \qquad (3.39)$$

Since τ is constant with height, r_{aM} varies little if it refers to wind well away from the surface where the wind gradient is small. Following Jarvis *et al.* (1976) canopy height (h) is taken as the reference level and (3.39) is written in the form

$$\frac{u_*}{u(h)} = \frac{k}{u[(h-d)/z_o]} = \eta_*. \qquad (3.40)$$

Substituting this into (3.39), with $u(z) = u(h)$ gives (3.41)

$$g_{aM} = 1/r_{aM} = C_{aM} u(h) = \eta_*^2 u(h) \qquad (3.41)$$

where C_{aM} is a dimensionless drag coefficient reflecting the properties of the canopy surface:

$$C_{aM} = [u_*/u(h)]^2 = \eta_*^2. \qquad (3.42)$$

Jarvis *et al.* collated values of η_* for coniferous forests and, although the range was wide, most fell within the range 0.25 to 0.35, leading to the convenient approximation that $C_{aM} \approx 0.1$ hence $r_{aM} \approx 10/u(h)$.

No similar data are available for deciduous forests but in view of the dependence of these parameter values on the values of z_o and d, and the uncertainty and variation in them, the above approximations will probably suffice for most purposes, particularly if wind speeds averaged over periods of more than an hour are being used in calculations. (This is not to say that there is no need for continued documentation of the relationships between the physical properties of canopies and their aerodynamic characteristics.) The caveats regarding the assumption that exchange coefficients are similar, expressed earlier, clearly apply to r_{aM}, but the above remarks also apply—there is little purpose in seeking precise correction factors to parameters which are only roughly known for particular canopies and which may be derived from fluctuating environmental variables averaged over arbitrary periods.

3.3.3 Transfer Inside Canopies

Considerable effort has been expended in attempting to apply one-dimensional exchange theory inside plant canopies, but it is becoming more and more obvious that, except as a very rough approximation, this is a futile exercise because the essential assumption—that air flow at any level is homogeneous in the horizontal—is not true. Nevertheless, it is worth considering briefly the factors affecting air flow in plant communities.

The momentum transmitted downward to a plant community—the shear stress (equation (3.36))—must be absorbed by the elements (stems and foliage) of the plant community, or the ground. This is described by the equation (due to Thom, 1971)

$$\tau(h) = \tfrac{1}{2}\rho_a \int_0^h u(z)^2 [a_f(z) C_{Mf}(z) + a_s(z) C_{Ms}(z)] \, dz \qquad (3.43)$$

where $a_f(z)$ and $a_s(z)$ are the foliage surface area and stem surface area per unit volume at level (z), and $C_{Mf}(z)$ and $C_{Ms}(z)$ are the corresponding effective values, within the canopy, of the individual leaf and stem drag coefficients. Equation (3.43) illustrates clearly the interaction between canopy structure and air flow within the canopy. The analysis of wind profiles in plant canopies using this approach was carried through by Landsberg and Jarvis (1973) in an attempt to determine exchange coefficients in a Sitka spruce canopy. They used an equation similar to (3.37) to determine the exchange coefficients. However, the analysis is fraught with problems, such as those of the drag coefficients and shelter factors (mutual aerodynamic interference between plant parts—see Landsberg and Thom (1971), Landsberg and Powell (1973)). It provides interesting insights into air flow in vegetation but the quantitative results are too inaccurate to be of general value.

Current thinking about exchange processes in canopies tends to the view that it is a rather intermittent process. There is continuous exchange between leaves and air inside the canopy but bulk exchange only at intervals when a (large) eddy or pressure pulse, with the properties of the above-canopy air, penetrates the canopy and changes the properties of the canopy air. There is then an interval during which the canopy air tends towards a condition depending on leaf energy balance and exchange properties and local air flow,

3 Stand Structure and Microclimate

before the next "flushing". Tajchman (1981) and Denmead (1984) have described this process but no mathematical model has yet been published.

Although rigorous analysis of bulk exchange processes inside plant communities is not currently feasible, a description of air movements in plant canopies is essential for analysis of leaf to air transfer processes. Wind profiles in canopies can be described by two empirical expressions:

$$u(z) = u(h) \exp[-n_u(1-z/h)] \qquad (3.44)$$

and

$$u(z) = u(h)[1 + m_u(1-z/h)]^{-2}. \qquad (3.45)$$

Equation (3.44) is due to Cowan (1968); (3.45) was given by Landsberg and James (1971) and by Thom (1971), n_u and m_u are numerical parameters characteristic of vegetation type. Values of $n_u = m_u = 3$ produce "reasonable" profiles. Given $u(z)$ and information about the dependence of r_b on wind speed (see Landsberg and Thom, 1971; Jarvis et al., 1976), appropriate values of r_b can be calculated for each level in a canopy. Now given an estimate of $\varphi_{nf}(z)$ from an equation such as (3.28), and the other necessary information on canopy microclimate (profiles of temperature and humidity), it is possible to calculate the flux of water vapour from leaves into the bulk air in the canopy. These calculations would normally be made over periods of about an hour. They are pertinent to the analysis of tree–water relations.

It is worth re-emphasizing here the point made earlier, namely that there is a close coupling between foliage energy balance and conditions within the canopy. Foliage temperatures depend on φ_{nf}, air temperature and humidity, stomatal resistance and wind speed. The flux of heat and water vapour from the leaves alters the conditions of the air within the canopy. In a dense canopy where bulk exchange processes are slow, the result of the feedback may be a quite marked modification of conditions inside the canopy. In open canopies with effective turbulent exchange around the leaves, the fluxes from the leaves do not significantly influence their ambient conditions, which remain dominated by the properties of the overlying air.

3.4 TRANSPIRATION FROM FOREST COMMUNITIES

Transpiration is one of the primary processes by which water is lost from forests, although it is difficult to generalize about the relative importance of the process in relation to other terms in the hydrological balance (see §7.5.1). In some climates evaporation of intercepted water may be much greater than transpiration losses, but transpiration remains of major importance because of its effects on plant water status. In this section I examine briefly the formulations for the calculation of transpiration which emerge from the preceding consideration of forest microclimate.

The rate of transpiration by a forest, or any plant canopy, depends on interactions between a number of variables, some of which are the properties of the environment and some of the plants. The properties of the environment have been discussed in this chapter; the role of the plant properties emerged in consideration of the leaf energy balance (equation (3.31)) and the role of stomata in determining the partitioning of energy. Clearly, we can estimate water loss from forests (transpiration plus evaporation from the soil and/or wet foliage) from measurements of φ_n and the Bowen ratio (see

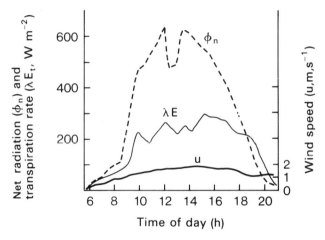

Fig. 3.12. Diurnal variation in net radiation (φ_n), transpiration rate and wind speed (u) for a stand of Douglas fir (*Pseudotsuga menziesii*) in coastal British Columbia on a day of intermittent cloud. Mean hourly values of λE were measured by the Bowen ratio method (redrawn from McNaughton and Jarvis, 1983).

Fig. 3.12). Given estimates of φ_s from sunshine hours (equation (2.8))—or measured values of φ_s—estimates of φ_n (e.g. from equation (3.15)) may be acceptably accurate. The use of a daily average value for β is likely to be less accurate but, averaged over long periods, this may well be the best way of estimating λE_t for a canopy. It is worth noting, in this respect, that

Landsberg (1984) estimated monthly average values of β by assuming a relationship with monthly rainfall. The results appeared "reasonable" but were untested. However, they suggest that evaluation of long-term empirical estimates of β, for different forest types and climates, may be useful.

In some circumstances we might assume that leaf (T_f) and air temperatures (T_a) are equal, then $(e_s(T_f) - e_a) \approx D$, the vapour pressure deficit. We can then, in principle, estimate the transpiration rate of a forest if we have a value for the canopy conductance (g_c). Canopy conductance depends on stomatal conductance and can be estimated by summation of the product of leaf area and stomatal conductance in each layer i, i.e.

$$g_c = \sum_{i=1}^{n} g_{s.i} \cdot a_{f.i} \qquad (3.46)$$

Average values of stomatal conductance for each layer $(g_{s.i})$ may be obtained either by measurements using porometers—in which case there

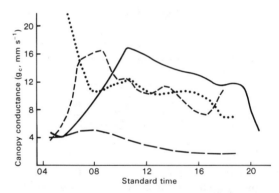

Fig. 3.13. Diurnal variation in canopy conductance for four conifer species: ——— *Pinus radiata*, derived using transpiration rates measured by weighing lysimeters; ——— *Pseudotsuga menziesii*, derived using transpiration rates measured by flux gradient techniques; ···· *Picea sitchensis*, derived using transpiration rates measured by flux gradient techniques: —— *Pinus sylvestris*, derived from porometer and leaf area measurements (see equation (3.46)). The data were collated from several sources by Whitehead (1985).

are severe sampling problems, or estimated from models (see §4.2). Figure 3.13 shows the diurnal course of canopy conductance values obtained by a number of investigators for several species.

The best known and most widely used equation for calculating transpiration from plant communities is the Penman–Monteith equation. This is a combination (energy balance–mass transfer) equation that essentially considers the canopy as a single large leaf. It can be derived from the leaf energy balance equation. For clarity here this will be rewritten using conductances rather than resistances:

$$\varphi_n = \rho_a c_p (T_f - T_a) g_H + \lambda E_t \qquad (3.47)$$

$$\lambda E_t = (\rho_a c_p/\gamma)(e_s(T_f) - e_a) g_V \qquad (3.48)$$

where

$$g_V = 1/(r_{bV} + r_s) \quad \text{and} \quad g_H = 1/r_{bH}.$$

Leaf temperature, and hence $e_s(T_f)$, can be estimated from the approximation

$$e_s(T_f) - e_a = [e_s(T_f) - e_s(T_a)] + [e_s(T_a) - e_a]$$

$$\approx (T_f - T_a)s_* + D \qquad (3.49)$$

where s_* is the slope at T_a of the saturation vapour pressure/temperature curve. Substituting (3.48) into (3.47) and combining with (3.49) gives

$$\lambda E_t = \frac{s_* \varphi_n + \rho_a c_p g_H D}{s_* + \gamma(g_H/g_V)}. \qquad (3.50)$$

Equation (3.50) becomes applicable to canopies if we replace the vapour conductance term g_V by canopy conductance g_c (from equation (3.46)), and replace g_H by a canopy boundary layer conductance g_a. Values for g_a may be estimated by assuming $g_a \approx g_{aM}$ (equation (3.39)), i.e. $g_a \approx u_h/10$ (m s^{-1}).

The Penman–Monteith equation provides a rigorous method of combining energy balance, aerodynamic and mass transfer parameters. It is widely accepted and has been found, by comparison with isotopic methods (Waring et al., 1980; Luvall and Murphy, 1982), and soil–water measurements and lysimetric data (Calder, 1977, 1978) to give good estimates of the transpiration rates of forests. Figure 3.14, from Waring et al. (1980) shows the relationship between cumulative daily values of E_t—calculated using equation (3.50) and measured by monitoring the movement of ^{32}P. There have been numerous studies of canopy resistance values (e.g. Stewart and Thom, 1973; McNaughton and Black, 1973; see also Fig. 3.12) but it must be recognized that the model ignores differences in the level of sources of heat and water vapour in the canopy and that there is little evidence (or reason to

3 Stand Structure and Microclimate

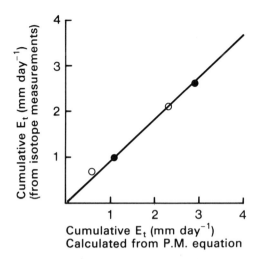

Fig. 3.14. Relationship between cumulative daily transpiration estimates made with the Penman–Monteith equation (equation (3.50)) and an isotope injection technique which allows calculation of the flux of water per unit sapwood basal area. The full and open circles represent data from two different plots of Scots pine (*Pinus sylvestris*) (redrawn from Waring *et al.*, 1979).

suppose) that the canopy conductance term can be rigorously identified with the stomatal conductance of all the leaf surfaces acting in parallel.

For conifers or other small-leaved trees, particularly in aerodynamically rough canopies where turbulent exchange is very effective, the ratio g_a/g_c may be large so that the right-hand term of (the canopy version of) equation (3.50) is large relative to the left-hand (radiation) term. The equation may then be simplified to

$$\lambda E_t = (\rho_a c_p/\gamma) D g_c \tag{3.51}$$

(see McNaughton and Black, 1973; Tan *et al.*, 1978). Equation (3.51) may be seriously in error for high-energy environments, days with low wind speeds or large-leaved plants. It has not been well tested.

A problem with equation (3.50) arising from its derivation as a "big-leaf" approximation, is that it ignores the soil. If the φ_n term is corrected for changes in heat storage in the soil and biomass ($\varphi_n - G$), then the equation applies strictly to the leaves, but even where this correction is made, errors arise because of the fluxes of water vapour to and from the soil surface and their effects on canopy microclimate. A solution to this problem would be to use a complex model incorporating soil surface properties (e.g. Norman, 1979), but for more general longer-term use, and for canopies with low L^*, it may be better to use the equation on a leaf area basis, ignoring the soil if it can be assumed to be dry and including corrections for periods when the soil is wet (see Denmead, 1984).

3.5 CONCLUDING REMARKS

The transient nature of microclimatic conditions in a canopy may seem to render such short-term phenomena irrelevant on the time scale of forest growth. However, as is the case with many processes, a sound understanding of the mechanisms involved is an essential basis for simple, longer-term analyses. We may illustrate this by considering the relationships between stand structure, energy interception and stand productivity.

Anticipating the more detailed discussion in Chapter 8, we note that it has been well established for agricultural crops that net dry matter increment of a community (ΔW) over a period of time Δt is proportional to radiant energy absorbed by the community (φ_{abs}) (Monteith, 1981; Jarvis and Leverenz, 1983), i.e. $\Delta W \propto \varepsilon_\varphi \varphi_{abs}$, where ε_φ is the energy conversion efficiency factor. This relationship has been shown to be applicable to forests (Linder, 1985)— using equation (3.28) with the incident energy (φ_s) term integrated over the periods between sequential harvests of plantation trees. The relationship could also be established by simulation, using the detailed light interception equations (3.23)–(3.26) and information on the photosynthetic properties of leaves (see Chapters 4 and 5). Differences in stand structure have important implications in relation to water use by understory plants, which in plantation forestry are regarded as weeds. Weeds usually thrive during the early stages of stand growth, when the canopy is incomplete. At this time weed cover may have serious effects on stand water balance, particularly in regions where water limits growth. The probable influence of weeds, in this respect, can be evaluated by calculating stand water balances, using equation (3.29) to calculate the amount of radiant energy incident on the weed cover. Applying estimates of the Bowen ratio (equations (3.17), (3.22))—obtainable from the agronomic or ecological literature (see Monteith (1973), pp. 193–199 (inclusive) for useful guidance) will provide an estimate of the amount of water likely to be used by weeds. Such figures are only estimates, and should be evaluated by measurements at several levels—e.g. stand water balance, stomatal resistance and leaf area of the weeds, water relations of weeds and trees (see Chapter 7) and microclimatic measurements. Similarly, the probable influence of mulch and bare soil surfaces under incomplete canopies can be evaluated by surface energy balance measurements.

In commercial forestry practice stand structure is often manipulated by thinning. Stands may be thinned "from below", when the smaller, poor quality trees are removed, or "from above", when the dominant trees are removed. Thinning "from below" will reduce canopy density and competition for water and nutrients. It is unlikely to improve the light regime much as this will largely be determined by the dominant trees. Thinning "from above" is likely to improve the light regime as well as reducing competition for water and nutrients.

The effects of thinning on stand structure can be evaluated using equations such as (3.2) to (3.4) and the likely effects on energy interception evaluated using equation (3.28). Energy interception and photosynthesis models, applied to thinned stands, allow estimates of the effects of thinning on stand growth rates and hence of the individuals. If the $W/\varphi_{s.abs}$ relationship discussed earlier has been established it can also be used to evaluate the early effects of thinning on stand growth rate.

Foresters tend to speak of thinning in terms of the amount or proportion of basal (trunk) area to be removed from a stand, and base their predictions of the responses they expect from the procedure on empirical experiments, which may or may not cover the conditions likely to be experienced at the site in question. Such predictions can undoubtedly be improved by analysing the results of thinning experiments in terms of the effects of the treatments on canopy structure, and hence on microclimate, particularly in relation to energy interception and dry matter production by the remaining trees. Again, knowledge of the mechanisms underlying empirical results allows safer extrapolation.

The equations for calculating stand transpiration, particularly the combination equation (3.50), are vital for estimating stand water balances and rate of water movement through trees (see Chapter 7). As well as its use in predicting transpiration rates, equation (3.50) provides a convenient analytical framework, within which the effects of leaf area and stomatal conductance and its dependence on environmental conditions (see §4.2) can be examined, in relation to the aerodynamic characteristics of stands and their energy balance.

In general, knowledge about the interactions between stand structure and microclimate is invaluable in analysing the reasons for differences in forest growth on different sites and in site assessment. Sites are usually assessed for forestry in terms of site indexes of various sorts. There is enormous potential in the use of mechanistic models to evaluate potential site productivity. This idea is discussed in more detail in Chapter 8.

4 The Carbon Balance of Leaves

The rate of dry matter production by forests—or any plant community—depends on the interception of radiant energy by leaves and the conversion of this energy into carbohydrate. The proportion of incident energy intercepted and absorbed by forest canopies depends, as discussed in the previous chapter, on the leaf area of the stand and the way the foliage is distributed. The effectiveness with which radiant energy is converted to chemical forms depends on the photosynthetic properties of the leaves.

Forest productivity may be described as the net rate of dry matter production, or the carbon balance of a stand, over any specified period. It depends on the rate of carbon fixation relative to the rate of loss by respiration, the death of individual trees and the death and shedding of organs. Commercial forestry is concerned with optimizing forest productivity, and ensuring that the maximum possible amount of accumulated carbohydrate is converted into marketable produce. All silvicultural practices are, in effect, directed towards these ends, although—except for the attention paid to tree form in breeding and selection programmes—most management practices are concerned only with increasing productivity (defining productivity more narrowly—and less accurately—in terms of the rate of production of the marketable product). Practices such as thinning are, at least implicitly, aimed at producing the optimum canopy size and structure; fertilization increases leaf area and may improve the photosynthetic efficiency of leaves.

Analysis of the carbon balance of trees and forest stands must begin with analysis of the carbon balance of leaves. This involves consideration of photosynthesis, respiration and carbon export. It also requires that we examine the question of stomatal conductance, since stomata provide the pathways through which CO_2 diffuses into leaves and to the mesophyll cell walls.

Exposing wet mesophyll cell walls to absorb CO_2 inevitably results in loss of water (transpiration) through the same stomatal pathways. The stomatal pores introduce resistance to the diffusion of water molecules into the atmosphere and constitute a major barrier to CO_2 assimilation. This barrier becomes progressively more significant when stomatal conductance is reduced by factors such as water stress. In this chapter, consistent with the pattern of this book, I will confine consideration to the empirical relationships between the stomatal conductance of leaves (g_s), environmental factors, and leaf condition. We will not consider the mechanisms of stomatal response at the cellular level, but some detail at the biochemical level is essential for understanding photosynthesis, the factors affecting it and how these factors operate. Such background is also essential as a basis for the use and interpretation of empirical equations.

4.1 PHOTOSYNTHESIS

Photosynthesis comprises light and dark reactions that involve the removal of electrons from water—resulting in the release of O_2—and the donation of these electrons to CO_2, leading to reduced carbon compounds with a gain in free energy. The process takes place in the chloroplasts. The basic equation may be written

$$H_2O + CO_2 \xrightarrow{energy} O_2 + (CH_2O).$$

The primary photochemical processes take place when light energy is absorbed by the photosynthetic pigments, which raises the energy level of the light harvesting chlorophyll molecules to an excited state. A specialized chlorophyll molecule donates electrons to electron carriers, which then flow down the electron transport chain. The energy of the electrons is used to generate ATP (adenosine triphosphate) and $NADPH_2$ (nicotinamide adenine dinucleotide phosphate (reduced)) (see Fig. 4.1). The photophysical and photochemical light reactions proceed at a rate that depends on the quality and intensity of light alone. These reactions are not affected by temperature or CO_2 concentration. Electron transport is strongly dependent on temperature.

The dark reactions use the energy (ATP) and reducing power ($NADPH_2$) produced in the light reactions to reduce CO_2 to carbohydrate (CH_2O). The initial acceptor of CO_2 is ribulose-bisphosphate (RuBP), the reaction being catalysed by the enzyme RuBP carboxylase. The first carbon reduction product in most trees is a 3-carbon (C_3) compound, 3-phosphoglyceric acid (PGA), reduced by ATP and $NADPH_2$, which further metabolizes to form sugars. The RuBP is regenerated in the Calvin–Benson cycle. (There is a large and important group of plants—including some trees—where the first carbon reduction product is a C_4 compound. However, these will not be considered in detail here.) RuBP carboxylase constitutes a major fraction of

4 The Carbon Balance of Leaves

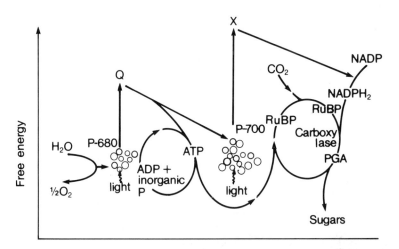

Fig. 4.1. Diagrammatic representation of the primary processes in photosynthesis. Light energy absorbed by photosystem II (PS-II), chlorophyll removes an electron from water, resulting in the evolution of oxygen. Energy is transferred from the excited chlorophyll molecules to a (reduced) substance, Q. Electrons pass from Q to PS-I, some of the energy being used to generate ATP from ADP. Light harvesting chlorophyll in PS-I provide the energy for substance X to reduce NADP—an electron carrier—to NADP H_2. Ribulose–bisphosphate (RuBP) is the initial acceptor of CO_2, the reaction being catalysed by RuBP carboxylase. RuBP is regenerated. The energy of ATP is used in the reactions in which CO_2 is reduced to PGA (=3-phosphoglyceric acid) and hence to sugars.

leaf protein but, because it has a relatively low affinity for CO_2 and is competitively inhibited by the oxygen fixing enzyme (oxygenase), it has been implicated as a factor that limits the rate of photosynthesis. The regeneration of RuBP appears to be dependent on the partial pressure of CO_2 at the carboxylation sites ($p(CO_2)$). If $p(CO_2)$ is low, CO_2 assimilation is not limited by the amount of RuBP but by the amount of the enzyme (RuBP carboxylase). As $p(CO_2)$ increases, electron transport reactions, and therefore the capacity to regenerate RuBP, become limiting, i.e. the supply of RuBP becomes limiting. A detailed mechanistic model of photosynthesis utilizing these principles was developed by Farquhar et al. (1980) and subsequently modified by Farquhar and von Caemmerer (1982). Figure 4.2 provides a summary of the limitations imposed on photosynthesis by diffusion of CO_2 to the sites of photosynthesis and by the carboxylating enzyme. The diagram shows the commonly observed relationship between CO_2 assimilation rates

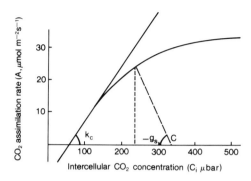

Fig. 4.2. CO_2 assimilation rate as a function of intercellular CO_2 concentration (C_i). The diagram is redrawn from Farquhar and Sharkey (1982), who designated it the "demand function" for CO_2; the broken line connecting A to ambient CO_2 (C_a) is the "supply function", with a slope of $-g_s$ (see equation (4.1)). The vertical (broken) projection from the A curve to the C_i-axis cuts that axis at the current value of C_i. The linear portion of the demand function curve is considered to be the region where the photosynthetic mechanism is RuBP saturated. The initial slope (dA/dC_i) gives the carboxylation efficiency (k_c) (see equation 4.11)).

(A) and the intercellular concentrations of CO_2 (C_i) (Wong et al. (1978) with *Eucalyptus pauciflora*; Watson et al. (1978) with apples; Pearcy et al. (1982) with Euphorbia species). The A against C_i curve has been designated the "demand function", while the line connecting C_a to A is the "supply function" (Raschke, cited by Farquhar and Sharkey, 1982). The slope of the supply function line is $-g_s$ (equation (4.1), below) and the downward projection from the point of intersection of the demand and supply functions to the C_i axis gives the value of C_i. The linear portion of the curve has been designated by Farquhar and Sharkey as RuBP saturated; in this part of the curve there is ample substrate and any increase in C_i results in activation of more enzyme, which increases the rate at which CO_2 is fixed. However, if the rate of RuBP carboxylation is increased sufficiently, the capacity to regenerate the substrate becomes limiting, and further increases in C_i do not lead to concomitant increases in A. This suggests that there is some optimum value of g_s that will lead to maximum photosynthesis for a particular leaf condition (e.g. nitrogen status).

This simplified outline of the biochemistry of photosynthesis ignores many uncertainties and sources of variation but will suffice for present purposes.

The equation describing CO_2 uptake as a process of diffusion into the leaf can be written

4 The Carbon Balance of Leaves

$$A = g_s(C_a - C_i)/P \qquad (4.1)$$

where C_a is the partial pressure of ambient CO_2 concentration (usually about 320 μbar), P is atmospheric pressure and

$$g_s = 1/(1.37 r_{bV} + 1.65 r_s) \qquad (4.2)$$

Boundary layer resistances (r_b) were discussed in Chapter 3; stomatal resistance (r_s) is discussed in §4.2. As the molecular diffusivities for water vapour and CO_2 in air are different, the r_b and r_s values used to calculate g_s for CO_2 must be corrected by the ratios of the diffusivities. The ratio is r_s (water): $r_s(CO_2) = 1:1.65$. The ratio for boundary layer resistances is 1.37 because of the influence of turbulence.

It has been widely assumed that $p(CO_2)$ at the chloroplasts (C_c) is near zero and hence that it is necessary to introduce another resistance in the pathway from C_i in the intercellular space to the chloroplasts. Jarvis (1971) describes how to calculate this. It is essentially the initial slope of the demand function (Fig. 4.1) and is therefore likely to reflect the effectiveness of enzymatic CO_2 fixation. Farquhar and von Caemmerer (1982) have recently suggested that chloroplast and intercellular CO_2 concentrations are effectively the same, hence this (so-called) mesophyll resistance can be ignored. This is not yet clearly demonstrated.

Equations (4.1) and (4.2), in association with Fig. 4.2 provide a great deal of information about leaf photosynthesis, and equation (4.1) also indicates that, if C_a were constant, leaf carbon assimilation rate can be expected to be linearly related to the conductance, g_s. This expectation has been realized in a number of studies (see Wong *et al.* (1979), and references cited there; also Fig. 4.3).

The linear relationship between A and C_i is discussed further in the next section, where the factors affecting stomatal conductance are considered.

4.2 STOMATAL CONDUCTANCE

The implication of the linear relationship between A and g_s—that C_i must be approximately constant—led Wong *et al.* (1978, 1979) to investigate the A/g_s relationship in detail. They proposed that stomata respond to changes in C_i, maintaining it at a constant level by negative feedback, i.e. stomatal aperture may be determined by the capacity of the mesophyll tissue to fix carbon and by some sort of communication from the photosynthetic mechanism to the stomatal control system.

This idea of feedback control is not universally accepted, but there are

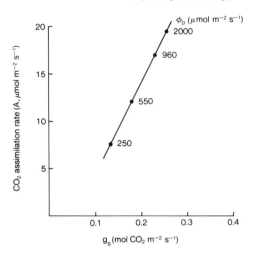

Fig. 4.3. Data from Wong et al. (1979) for *Eucalyptus pauciflora*, showing the linear relationship between assimilation rate and conductance to CO_2 (g_s), measured at four irradiances, with ambient conditions held constant.

many data showing that leaf conductance and rate of assimilation increase with irradiance in very similar fashion (see Burrows and Milthorpe, 1976). Furthermore, Schulze and Hall (1982) present highly significant linear regressions between maximal (with respect to φ_p) CO_2 assimilation rates (A_{max}) and concurrent leaf conductance values. These regressions were relatively constant for plants of the same species, but varied between plants of different species. Schulze and Hall conclude that, in natural environments, the range of stomatal conductance (g_s) varies so that it (apparently) matches the photosynthetic capacity of leaves as this capacity is determined by the long-term effects of several environmental and plant factors. This conclusion is consistent with the Wong et al. (1978) hypothesis of feedback between C_i and g_s, resulting in constant C_i.

The feedback hypothesis is of great interest, and will undoubtedly stimulate considerable research, but it is as yet of limited use as a means for predicting the values of g_c required for calculating the carbon and water balances of leaves. For this purpose we must resort to more empirical models, such as those discussed in the next section where the relationships between g_s, irradiance and vapour pressure deficit are considered. Temperature also affects g_s, but this factor has been neglected as being usually second order and difficult to account for in varying (natural) environments.

We should also note that, in dynamic situations, for example with rapidly changing φ_p (see Fig. 2.3) and changing vapour pressure deficit, C_i is unlikely to be constant. However, the assumption of constancy—caused by feedback—is probably acceptable over periods of order 10^3 s.

4.2.1 Effects of Irradiance and Vapour Pressure Deficit

Stomatal conductance is usually measured by determining the flux of water vapour per unit leaf surface area (E_f), either in a gas exchange system or using some form of diffusion porometer. In most of the literature on stomatal conductance the units used are m s^{-1} (or mm s^{-1}). Where flux density is expressed in molar terms (as is now usual) g_s has units mol m^{-2} s^{-1} (see Jones, 1983b). Therefore, from flux measurements,

$$\frac{1}{g_s} = \frac{e_i - e_a}{E_f P} - \frac{1}{g_{bV}} \tag{4.3}$$

where e_i is the vapour pressure at leaf temperature, e_a is ambient vapour pressure, P is atmospheric pressure, and g_{bV} is boundary layer conductance.

Jarvis (1976) presented a model to describe the response of stomata to environmental variables and applied it to temperate conifers. He used results from controlled environment studies to choose the empirical function which best described the response of stomatal conductance to each environmental variable (see Fig. 4.4.). Non-linear least squares techniques were then used to fit these functions to conductance measurements made in the field, where all the environmental variables were also measured. Thorpe et al. (1980) analysed data on the responses of apple leaf stomata to environmental factors and, guided by the relationships they observed, developed a model for single leaves and a whole tree. Whitehead et al. (1981) measured stomatal responses to environmental variables in the field in two tropical forest species in Nigeria and followed Jarvis's approach in fitting the conductance values to environmental variables. From these, and many other studies, it is clear that the stomata of trees respond to environmental variables in the same way as those of most other plants with normal between-species (and between-individual) variation. The main controlling factors are light (photon-flux density, φ_p) and air humidity. Water stress does not become a factor until leaf water potential (ψ_f) falls quite low (see Landsberg et al., 1976; Beadle et al., 1978) and the effects of varying CO_2 concentrations need not be included in empirical models for plants well coupled to the environment.

To illustrate the diurnal patterns of stomatal conductance, Fig. 4.5 shows diurnal conductance patterns observed by Watts et al. (1976) in Sitka spruce

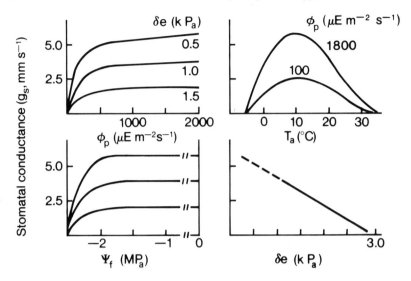

Fig. 4.4 Response functions from Jarvis' (1976) model of the dependence of stomatal conductance on environmental factors. The functions were indicated by measurements on Sitka spruce (*Picea sitchensis*) and Douglas fir (*Pseudotsuga menziesii*) and, although the form may suit other species, the parameter values would almost certainly be different (see Fig. 4.6 for a more detailed function for ψ_f).

in Scotland and Whitehead *et al.* (1981) in teak in Nigeria. Species of eucalypt studied by Sinclair (1980) showed diurnal patterns similar to this.

The models developed by Whitehead *et al.* and Thorpe *et al.* are not dissimilar. Both are of the form

$$g_s = f(\varphi_p, D) \tag{4.4}$$

(where D is vapour pressure deficit) and, since the experimental results indicated that φ_p and D act independently, both were rewritten as

$$g_s = f_1(\varphi_p) f_2(D). \tag{4.5}$$

Also in both cases the response to photon flux was a hyperbolic increase in g_s with increasing φ_p, while the response to vapour pressure deficit was a linear decrease in g_s with increasing D.

Thorpe *et al.* (1980) expressed their model as a single equation, although it is necessary to determine parameter values for different species and to test

4 The Carbon Balance of Leaves

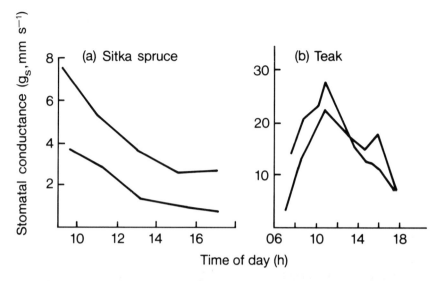

Fig. 4.5. Diurnal patterns of stomatal conductance in (a) Sitka spruce (*Picea sitchensis*) (from Watts *et al.*, 1976) and (b) Teak (*Tectona grandis*) (from Whitehead *et al.*, 1981). The spruce data were obtained at two levels in the canopy, the upper curve being from higher in the canopy. The teak curves were from two different days. Note the completely different diurnal trends and the (approximately) 4-fold difference in maximum g_s. Watts *et al.* attributed the decrease in g_s mainly to increasing leaf–air vapour pressure difference (δe) (see Fig. 4.4). Whitehead *et al.* found that g_s was responding to several factors, with φ_p and δe dominant.

that the responses implied in the model will hold where it is to be used. The equation of the model is

$$g_s = g_{ref}(1 - aD)/(1 + b/\varphi_p) \qquad (4.6)$$

where a and b are empirical "constants" and g_{ref} is a reference conductance. The parameter values may be determined from measurements of stomatal response to φ_p at low values of D (say from 0.5 to 1.0 kPa) and responses to D when φ_p is not a limiting variable. In both cases the analysis uses values of g_s normalized to the highest observed value; i.e. that value is taken as unity. This greatly reduces the scatter in data. The reference value (g_{ref}) is then the value of g_s expected when both D and φ_p are non-limiting, i.e. it is the maximum value of g_s. Körner *et al.* (1979) list maximum leaf conductance values for 294 species. For most woody species the values range from 1 to 5 mm s^{-1}. If we take $a=0.3$ kPa^{-1} and $b=70$ µmol m^{-2} s^{-1}, then $g_s = 0$ when $D=3$ kPa and $g_s = 0.5 g_{ref}$ when $\varphi_p = 70$ µmol m^{-2} s^{-1}.

The consequences of stomatal response to humidity in terms of transpiration rates have been explored by Landsberg and Butler (1980). They showed that as stomata close in response to increasing vapour pressure deficit and the proportion of energy dissipated as latent heat (see equation (3.31)) is reduced, boundary layer resistance exerts an increasingly important effect on transpiration rates. As a result leaf temperature—and hence $(T_f - T_a)$—increases. Consequently, for plants with stomata sensitive to humidity, the rate of transpiration in a high-energy environment can be similar to that in a low-energy environment.

4.2.2 Effects of Water Stress

Stomata generally appear insensitive to leaf water potential until some threshold (critical value) of ψ_f, below which stomatal closure takes place, is reached. Schulze and Hall (1982) consider that the evidence for this is not particularly good, but it has to be accepted that stomata close with drought. The point at which closure occurs varies considerably, depending on factors such as leaf age and pre-treatment (e.g. acclimation to drought). In general, closure is likely to occur at ψ_f values of about -2 MPa. A thorough study by Beadle et al. (1978) on the critical values of ψ_f (ψ_{crit}) for stomatal closure in Sitka spruce (*Picea sitchensis*) gave values between -2.5 in the upper canopy (high transpiration rates) and -1.8 MPa in the lower canopy (shaded; low transpiration rates). Values of ψ_{crit} for seedlings grown under different irradiance regimes were all between -2 and -2.5 MPa.

The general effect of leaf water potential on stomata is illustrated in Fig. 4.6. The relationship can be described by a curve, e.g. $g_s = g_{ref}/[1+(\psi_f/\psi_{crit})^n]$, where n would have a value of about 2 or 3. Alternatively two straight lines may be used, i.e. if $\psi_f > \psi_{crit}$, $g_s = g_{ref}$, and if $\psi_f < \psi_{crit}$ then $g_s = g_{ref} + m\psi_f$ (where the correct (negative) sign is used for ψ_f).

We may therefore modify equation (4.6) to

$$g_s = g_{ref} \frac{(1-aD)}{(1+b/\varphi_p)} f(\psi_f) \qquad (4.7)$$

where $f(\psi_f)$ denotes a relationship to describe the effects of water potential.

The above discussion refers essentially to short-term water deficits. Leaves subjected to long periods of such deficits may acclimate to them, so that the

4 The Carbon Balance of Leaves

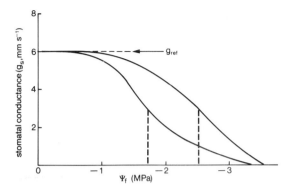

Fig. 4.6. The response function for stomatal conductance (g_s) in relation to ψ_f. The curves are described by the function $g_s = g_{ref}/[1+(\psi_f/\psi_{crit})^n]$, so when $\psi_f = \psi_{crit}$, $g_s = 0.5 g_{ref}$. The two curves were calculated using $n=2$ and different values of ψ_{crit}, indicated by vertical broken lines. The rate of reduction in g_s with ψ_f near ψ_{crit} depends on the value of n.

ψ_{crit} value falls much lower than in leaves normally subject to only transient stress. (See comment in Chapter 7, on effects of water stress.)

4.3 AN EMPIRICAL MODEL OF PHOTOSYNTHESIS

Gas analysis techniques are widely used for the study of photosynthesis and evaluation of the photosynthetic properties of leaves. These studies have led to the development of empirical models of the gas exchange of leaves—models that are of great value for evaluating the consequences of variations in the leaf photosynthetic properties, as described by the model parameters. Clearly these parameters should relate directly to biological processes, and models based on such processes are more valuable than those which are merely statistical descriptions of data.

The most comprehensive extant treatment of the biochemical and biophysical processes involved in CO_2 assimilation by a single leaf is that presented by Farquhar and von Caemmerer (1982). However, at time of writing, this was still in a form difficult to use in practice and requires both a great deal of knowledge and understanding of basic processes, and precise, detailed information on the photosynthetic properties of leaves—very little of which is available for forest trees. For the practical purpose of simulating the carbon balance of leaves, and subsequently of whole trees and communities of trees, it seems more useful to use a well-established empirical model for which parameter values are readily available. Nevertheless, we must recognize the shortcomings of such a model and the fact that it should, in due course, be supplanted by better models.

The mathematics of empirical models have been treated in detail by Charles-Edwards (1981). His basic development starts from the biochemical processes but the approach is quite different from that of Farquhar and von Caemmerer and is, as he says himself, essentially empirical. Watson et al. (1978) used a model similar to that which emerges from Charles-Edward's (1981) analysis. This is presented here.

Although it may not always provide the best description of particular data sets, the response curve of net photosynthesis to photon-flux density can often be described by the widely used empirical relationship

$$A = \frac{\alpha_p \varphi_p A_{max}}{\alpha_p \varphi_p + A_{max}} - R_d \tag{4.8}$$

where α_p is the quantum yield (efficiency) of photosynthesis, given by the initial slope of the CO_2 uptake/light response curve.

The light saturated assimilation rate A_{max} is given by

$$A_{max} = k_c(C_i - \Gamma) \tag{4.9}$$

where k_c is strictly the carboxylation efficiency (see Farquhar and Sharkey, 1982), although Watson et al., consistent with the assumptions of that generation of models, called it mesophyll conductance. Γ denotes the rate of CO_2 efflux in the light at zero C_i observed in C_3 plants. For practical empirical purposes this term can be ignored. Substituting for A_{max} in equation (4.8) the model becomes

$$A = \frac{\alpha_p \varphi_p k_c C_i}{\alpha_p \varphi_p + k_c C_i} - R_d \tag{4.10}$$

We note that equation (4.10) implies that the rate of CO_2 uptake increases hyperbolically with photon-flux density, with A_{max} being set by the values of k_c and C_i. A typical curve produced by the equation is shown in Fig. 4.7. Differentiating equation (4.10) gives

$$\frac{dA}{dC_i} = \frac{k_c}{(1 + k_c C_i / \alpha_p \varphi_p)^2} \tag{4.11}$$

which implies that when $\varphi_p \gg k_c C_i / \alpha_p$ (i.e. at saturating photon-flux densities),

4 The Carbon Balance of Leaves

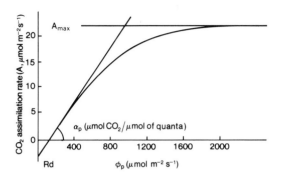

Fig. 4.7. "Standard" hyperbolic CO_2 uptake response to irradiance. The curve is of the form generated by equations (4.8) and (4.10). The parameters α_p, A_{max} and R_d are illustrated. (k_c is illustrated in Fig. 4.2.)

$dA/dC_i \approx k_c$. This provides an estimate of the carboxylation efficiency.

Reference to Fig. 4.2 indicates that the model is invalid at high values of C_i but it is likely to be adequate for C_i values associated with ambient $p(CO_2)$ of about 340 µmbar. There is therefore a need to experimentally establish A/C_i curves by using a wide range of C_a values to vary C_i. These will provide empirical values for k_c and information on the basic photosynthetic properties of leaves (see earlier discussion).

Using equation (4.1) to eliminate C_i from (4.10) leads to the quadratic

$$0 = A^2 + A[R_d - \alpha_p \varphi_p (1 + g_s/k_c) - g_s C_a] + \alpha_p \varphi_p g_s [C_a - R_d/k_c] - R_d C_a g_s$$

hence

$$A = B - (B^2 - C)^{1/2}$$

where

$$B = 0.5(R_d - \alpha_p \varphi_p (1 + g_s/k_c) - g_s C_a$$

and

$$C = \alpha_p \varphi_p g_s (C_a - R_d/k_c) - R_d C_a g_s.$$

In much earlier work irradiance was given in energy units (W m²) but to present them this way can be misleading because of the differences in photon-flux density when different types of lights of the same power output were used as the energy source in experiments. This problem may always be present in gas exchange work. In even older work, photometric units such as lux and foot candles were used, but results expressed in these terms should only be used (if at all) for approximate comparisons. With the ready availability of quantum sensors there is now little excuse for not using the correct units in photosynthetic research. However, in simulation work or analysis of field data, where irradiance data are usually expressed in energy units, linear conversions are adequate (for sunlight $1 \text{ W m}^{-2} \approx 2.3 \text{ }\mu\text{E m}^{-2}\text{s}^{-1}$).

4.3.1 Parameter Values and their Variation

Values for the parameters of equation (4.10) vary enormously and some obtainable from the literature are given in Table 4.1. They are discussed briefly here.

The theoretical quantum requirement for fixing CO_2 is 8 mol photons per mol of CO_2 fixed, i.e. the quantum yield α_p is 0.125 mol CO_2 fixed/mol photons. Ehleringer and Björkman (1977) in a study of the quantum yields of C_3 and C_4 plants in low O_2 or high CO_2 pressures found values in the range 0.073 to 0.081 mol CO_2/mol photons absorbed. In normal atmospheres the quantum yields of C_3 species are 0.05 to 0.054 mol CO_2/mol photons absorbed at 30°C. Because of the production of CO_2 by photorespiration they are strongly temperature dependent.

In most gas exchange systems it is not possible to determine how much of the energy incident on leaves is absorbed, so values of the quantum yield derived from such measurements are usually mols CO_2 fixed per mol of *incident* quanta. From a practical point of view, this may be no great disadvantage, since this is the information most easily available from light interception models or measurements. Nevertheless, we have to remember that the reflection coefficients of leaves vary and, in the case of conifers, the angle of incidence may vary enormously within an assimilation chamber in which irradiance is not totally diffuse. Furthermore, there is the problem, already mentioned, of quality of light sources and other problems associated with the engineering and technology of chambers. Also the number of photons absorbed by chloroplasts depends on chloroplast packing and concentration in leaves.

The higher values of α_p in Table 4.1 are those obtained in the laboratory, particularly where considerable care was taken to ensure optimum illumination of the leaves. Values obtained from field measurements indicate $\alpha_p \approx 0.03$ mol CO_2 per mol (incident) quanta; this might serve as a reasonable general value to use in calculating photosynthesis by forest trees.

Table 4.1. Published values or values derived from published data or graphs, of photosynthetic quantum yield (α_p), and rates of light saturated photosynthesis (A_{max}) of leaves of a number of tree species. For ease of comparison with many published data values are given in both molar and mass units.

Species	α_p		A_{max}		Comments	Source
	mol/mol	g/mol	µmol/m²/s	mg/m²/s		
Nothofagus solandri	0.032	1.41	10.0	4.4×10^{-2}	Sun leaf; field	Benecke and Nordmeyer (1982)
Pinus contorta	0.026	1.14	9.6	4.2×10^{-2}	Sun leaf; field	Benecke and Nordmeyer (1982)
Pinus radiata	0.026	1.14	15.8	6.9×10^{-2}	Sun leaf; field	Benecke (1980)
Pinus sylvestris	0.027	1.19	17.2	7.6×10^{-2}	Sun leaves; field	Troeng and Linder (1982)
Betula verrucosa	0.064	2.82	—	—	Light integrating sphere; laboratory	Brunes et al. (1980)
Pinus sylvestris	0.048	2.11	—	—	Light integrating sphere; laboratory	Brunes et al. (1980)
Picea sitchensis	0.061	2.68	9.3	4.1×10^{-2}	Light integrating sphere; laboratory	Ludlow and Jarvis (1971)
Agathis microstachya	0.018	0.79	5.0	2.2×10^{-2}	Sun leaf; laboratory	Langenheim et al. (1984)
Agathis robusta	0.024	1.06	7.3	3.2×10^{-2}	Sun leaf; laboratory	Langenheim et al. (1984)
Copaifera venezuelana	0.065	2.86	7.6	3.3×10^{-2}	Sun leaf; laboratory	Langenheim et al. (1984)
Hymenaea courbaril	0.056	2.46	5.0	2.2×10^{-2}	Sun leaf; laboratory	Langenheim et al. (1984)

We note, from equation (4.9), that A_{max} depends on the carboxylation efficiency k_c and internal CO_2 concentration C_i—in turn dependent upon stomatal conductance g_s (see equation (4.1)). Values of A_{max} may therefore be affected by stomata or reflect the nutritional status of leaves and the light regime in which they were grown. Farquhar and von Caemmerer (1982) calculate A_{max} values of 20–25 µmol m^{-2} s^{-1} for leaves not subject to serious limitations but the data in Table 4.1 suggest that such assimilation rates are not often achieved.

There are few values of the carboxylation efficiency (k_c) obtainable from the literature, but estimates can be obtained from graphs given in the papers by Ludlow and Jarvis (1971), who studied Sitka spruce (*Picea sitchensis*), and by Langenheim et al. (1984), who studied Amazonian rainforest species. For forest-grown spruce $k_c \approx 0.05$ µmol m^{-2} s^{-1} µbar^{-1}. Carboxylation efficiency was higher in sun leaves of both *Agathis microstachya* and *A. robusta* (0.025 and 0.04 µmol m^{-2} s^{-1} µbar^{-1} respectively) than in shade leaves (0.02 µmol m^{-2} s^{-1} µbar^{-1} in each case).

Leaf dark respiration varies approximately exponentially with temperature up to damaging temperatures; Watson et al. (1978) fitted an exponential equation to dark respiration data obtained from apple leaves. At about 20°C typical values are about 0.1 to 0.2 µmol m^{-2} s^{-1} (see also Langenheim et al., 1984).

4.4 CARBON CONSUMPTION AND EXPORT BY LEAVES

Leaf growth can be described in terms of the carbon balance of leaves, i.e.

$$dW_f/dt = C_{im} + Y(A - R_d) - C_{ex} \qquad (4.12)$$

where C_{im} denotes the rate of carbohydrate (assimilate) imported from elsewhere in the plant, Y is the conversion factor for CO_2 for carbohydrate and C_{ex} denotes the rate of export of carbohydrate. The leaf mass at any time is given by the integral of equation (4.12). In the early stages of the life cycle of a leaf, both A and C_{ex} will be zero and the leaf is essentially parasitic. When A becomes significant, C_{im} tends towards zero and active export of assimilate commences. The period of most active export occurs at about the time a leaf reaches its maximum size, although in Populus leaves at that time half YA was still being retained (Larson and Gordon, 1969). After maturity, export continues but the leaf photosynthetic capacity declines.

The time course of these processes will vary with species, leaf condition and the normal length of time leaves remain on trees. In deciduous trees the whole cycle takes place in a season. In trees which retain their leaves for a number of years, the leaves may export at a high rate for several seasons. Water stress tends to reduce the area of individual leaves, and their life span, and leaves grown in high light environments tend to be thicker (lower specific leaf area) than those grown in low light. Nitrogen nutrition in particular affects leaf size and specific leaf weight. All these factors interact.

Watson and Landsberg (1979) analysed the time course of leaf carbon balance in terms of changes in the value of the parameters of equation (4.11) and measured changes in leaf area and dry weight. This kind of analysis permitted them to develop relatively simple models of leaf growth and carbon balance and offers one approach to the study of growth patterns of forest trees.

4.5 CONCLUDING REMARKS

The A/C_i curve (Fig. 4.2) offers a valuable method of analysing the photosynthetic properties of leaves without the complications of diffusion through the stomates. A shift in the initial slope may reflect changes in the activity of the carboxylase while the maximum value of A (A_{max}), is limited by the regeneration of RuBP, and reflects leaf photosynthetic potential at a biochemical level. An example of the use of the method is provided by Pearcy *et al.* (1982), who used A/C_i curves to analyse the photosynthetic properties of 11 Hawaiian Euphorbia species. Among their findings was a high correlation between A_{max} and leaf nitrogen, that Pearcy *et al.* consider "probably reflects differing capacities for assimilation of N in leaf protein... A large fraction of leaf N is contained in the carboxylase enzymes in C_4 plants; (Björkman *et al.*, 1976); variations in leaf N may therefore reflect variations in the concentration of these and associated photosynthetic enzymes". (The fraction of leaf N in the carboxylase enzymes (primarily RuBP-carboxylase) is about 0.5, and the remark also applies to C_3 plants.)

Research on the photosynthetic properties of leaves, in terms of their gas exchange, necessarily involves measurements from which stomatal conductance can be calculated (equation (4.4)). However, for purposes such as calculation of photosynthetic rates from empirical formulae (or of transpiration rates, see equation (3.31)), models of stomatal conductance that describe how this parameter varies in relation to environmental conditions are required. Given such models (e.g. equations (4.6) and (4.7)) the rate of CO_2 uptake by leaves can be calculated from models incorporating terms related to some of the basic processes of photosynthesis, such as quantum yield and biochemical properties.

An important point that emerges from the rather cursory survey of values of k_c, α_p and A_{max} in this chapter is their wide variation. There is an urgent need for systematic studies of these photosynthetic properties of the leaves of forest trees, and the way they vary in relation to factors such as shade and nutrition. The models now available provide the opportunity to do this, and evaluate the extent to which variation in leaf photosynthetic properties is a consequence of genotype or environment. The way such knowledge could be used to assess the importance of these factors in stand productivity is discussed in Chapter 5.

5 The Carbon Balance of Trees

There are a great many data on forest biomass and the carbon balance of forest stands calculated from harvest data. There is little point in reviewing them here, particularly as Cannell (1982) has provided a comprehensive collation of forest biomass and primary production data; Satoo and Madgwick (1982) also provide a valuable data collection, with more analysis than Cannell.

However, to my knowledge, no serious attempt has yet been made to analyse forest productivity in terms of a model of the type suggested in this book—that is, a carbon balance model incorporating several organizational levels. The nearest approach to this was made by Jarvis (1981), who analysed the production efficiency of coniferous forests in the United Kingdom in terms of the processes involved in dry matter production. He compared the maximum rates of dry matter production of forest and agricultural crops and concluded that "despite low rates of photosynthesis, daily rates of CO_2 assimilation by conifer canopies are as large as by canopies of C_3 agricultural crops". However, Jarvis concluded that "we are remarkably ignorant about many of the factors which determine yield in coniferous crops". In a later, more detailed attempt, Jarvis and Leverenz (1983) analysed the main factors affecting the growth rate of stands—leaf photosynthetic properties and radiation interception, respiration and losses by root, leaf and tree mortality. Since an objective of this book is to set the analysis of forest growth rates within the framework of a quantitative process-based model we will examine the carbon balance of trees in similar terms (excluding tree mortality), with a view to moving towards the development of such a comprehensive model.

In commercial forestry estimates of standing biomass are normally obtained from empirical regression models; the standing mass or volume of timber is calculated from information such as tree number, height and diameter at breast height. Dynamic stand models depend on similar input information and may include growth curves, the probability of mortality, and tree size distribution functions. Estimates of growth patterns are of considerable value to managers, particularly if the model used was derived from data collected in the local area. However, they cannot provide answers to the "what if" type question: "What if there is drought or unusually high rainfall? What if thinning, or fertilization, are undertaken? What if there is insect attack or disease?" If the empirical base of a model is wide enough, it may provide useful practical answers to some of these questions. Even so, there is no flexibility to explore a wide range of possibilities, including the consequences of perturbing the system in a new way; e.g. by introducing a new genotype or management system or a new pest, disease, or problem such as acid rain. A model of the type envisaged in this book would provide this flexibility.

In this chapter, therefore, we will examine how leaf area and leaf photosynthetic properties can be coupled with light interception models to give estimates of tree and canopy dry matter production. Such estimates must include carbon losses by respiration. Estimates of respiration are provided without discussion but will be considered in more detail in a later section.

Total dry matter must be partitioned between roots, stems, branches and foliage. Hence, calculation of total dry matter production is only the first step in the process of estimating forest productivity and wood production. The relative sizes and growth rates of the component parts of trees vary greatly under different conditions, particularly in the amounts of assimilate allocated to roots, and we cannot assume constant ratios amongst parts.

Unfortunately, a mechanistic approach to calculating dry matter partitioning is precluded by our poor understanding of the mechanisms controlling carbohydrate partitioning. There are many data, obtained by destructive harvesting, on the ratios of the (aerial) component parts of trees to one another (allometric ratios, some of which are reviewed in this chapter), but very few studies of the mechanisms that lead to those ratios. Therefore, for some time to come, attempts to model carbon partitioning will depend on these empirical relationships.

Not surprisingly, there exists much less information on the variation in root mass and turnover relative to aerial parts of trees than on the relationships between aerial parts, but there has been considerable impetus to such studies in recent years and the information base is improving steadily. As a result it is now apparent that fine root turnover is a major source of carbon loss (expenditure might be a more appropriate word) for trees, so the factors affecting it must be identified.

5.1 CALCULATIONS OF CANOPY PHOTOSYNTHESIS

Leaf area, leaf photosynthetic characteristics, and photon-flux density, all have effects upon carbon assimilation by forests. To illustrate these effects, equations (3.23)–(3.26) have been used to calculate the diurnal course of canopy photosynthesis for two values of L^* for a day in spring and a day in summer in the southern hemisphere. The assumption was made that we were dealing with a continuous canopy and the two values of leaf area index were provided as inputs ($L^*=2$, $L^*=5$). Net CO_2 uptake was calculated hourly, the time course of photon-flux density being given by equations (2.6) and (2.7) ($\varphi_{p.max}$ was taken as 2000 µmol m^{-2} s^{-1} in February and 1500 µmol m^{-2} s^{-1} in October—see Fig. 5.2). It was assumed that the irradiance value for each hour is the value at the middle of the hour. In simulations intended to evaluate such a model in relation to measurements the hourly average measured value of photon-flux density would usually be used.

Solar zenith angles (equations (3.23), (3.26)) were entered in tabular form and the ratio of diffuse to total photosynthetically active radiation ($\varphi_{p.diff}/\varphi_p$) was assumed to be 0.4 in summer and 0.6 in spring.

Equation (4.10) was used to calculate photosynthesis; stomatal conductance was not needed in these calculations and C_i was taken as constant (=250 µbar). Two values of the parameters α_p and k_c were used; namely $\alpha_p=0.035$ and 0.015 µmol µmol^{-1} and $k_c=0.03$ and 0.015 µmol m^{-2} s^{-1} µbar^{-1} (cf. Chapter 4). Canopy photosynthesis is given by $A_{sun}L^*_{sun} + A_{shade}L^*_{shade}$ (see Fig. 5.1); assimilated CO_2 may be converted to dry matter equivalents by multiplying by 28.5 g mol^{-1}. (This is a general average value which may vary with the composition of dry matter—i.e. the proportions of carbohydrates, fats, proteins, etc.)

The curves in Fig. 5.2 show the expected effect of increasing L^* from 2 to 5. In this range the increase in total assimilation was non-linear—the ratio of total assimilation at $L^*=2$ to that at $L^*=5$ (averaged over both days and leaf photosynthetic parameters) was approximately 1.6. The values of α_p and k_c that have been used give maximum canopy assimilation rates (averaged over both days and L^* values) of about 13.3 µmol m^{-2} s^{-1} ($\alpha_p=0.035$ µmol µmol^{-1}, $k_c=0.03$ µmol m^{-2} s^{-1} µbar^{-1}), and 4.7 µmol m^{-2} s^{-1} ($\alpha_p=0.35$ µmol µmol^{-1}, $k_c=0.015$ µmol m^{-2} s^{-1} µbar^{-1}).

Total energy income was 64.2 mol for the 14 h summer day and 41.3 mol for the 12 h spring day. The diurnal uptake patterns in Fig. 5.2 therefore represent overall conversion efficiencies ranging from $10-12 \times 10^{-3}$ mol CO_2 per mol photons ($L^*=5$, $\alpha_p=0.035$ µmol µmol^{-1}, $k_c=0.03$ µmol m^{-2} s^{-1} µbar^{-1}) to $\sim 2 \times 10^{-3}$ mol CO_2 per mol photons ($L^*=2$, $\alpha_p=0.015$ µmol µmol^{-1}, $k_c=0.015$ µmol m^{-2} s^{-1} µbar^{-1}).

The implications of these calculations in terms of forest productivity are of considerable interest. To evaluate the productivity of a canopy over a season, the calculations should be made for each day. However, to get some idea of whether the calculations done here produce results that are reasonable, in

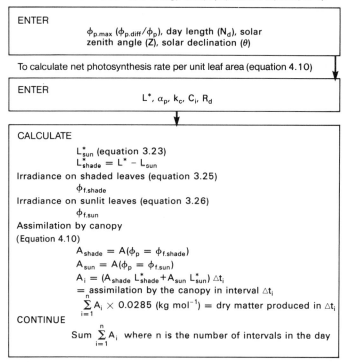

Fig. 5.1. Outline of the sequence of calculations done to produce Fig. 5.2. It is presented (roughly) in the form of a flow chart for a computer program; it will be clear how such programs—which may be much more complex than the one used here—would be written.

terms of forest productivity, we will consider the consequences of a single (hypothetical average seasonal) value of daily productivity.

Since tree population density is not specified these calculations of productivity per unit ground area tell us nothing about the growth rate of individual trees (dW/dt). This is given by the growth rate of the stand ($d(\Sigma W)/dt$ divided by the population per unit area (p). Hence, if p is large, dW/dt, for any $d(\Sigma W)/dt$, is small, and vice versa.

Calculated dry matter production $d(\Sigma W)/dt$ for the curves in Fig. 5.2 ranges from $3.1 \text{ g m}^{-2} \text{ day}^{-1}$ (October, $L^* = 2$) to $18.5 \text{ g m}^{-2} \text{ day}^{-1}$ (February, $L^* = 5$). Taking a representative value of say, 10 g dry matter m^{-2}

5 The Carbon Balance of Trees

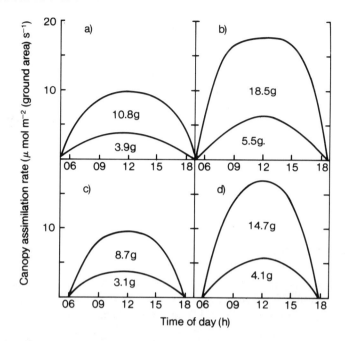

Fig. 5.2. Time curves of net assimilation by two hypothetical canopies in late summer (February curves a,b) and spring (October curves c,d). $L^*=2$ for curves a and c; $L^*=5$ for curves b and d. The calculations were made using Norman's (1982) radiation interception equations (3.23)–(3.26), and equation (4.10). The upper curve in each case is for a canopy with photosynthetically "efficient" leaves ($\alpha_p = 0.035$ μmol m^{-2} s^{-1} μbar^{-1}, $k_c = 0.03$); the lower curve is for less efficient leaves ($\alpha_p = 0.015$, $k_c = 0.015$). Max PAR values were 2000 μmol m^{-2} s^{-1} in summer and 1500 μmol m^{-2} s^{-1} in spring; peak photosynthesis rates were almost the same because 1500 μmol m^{-2} s^{-1} is nearly saturating. The numbers within the curves are canopy production in grams dry matter per day.

(100 kg ha^{-1}) day^{-1} we may partition this into leaf, root and stem growth: (dW_f/dt, dW_r/dt, dW_B/dt) for the stand as a whole:

$$\frac{dW_f}{dt} = \eta_f \frac{dW}{dt} \qquad (5.1a)$$

$$\frac{dW_r}{dt} = \eta_r \frac{dW}{dt} - \gamma_r \eta_r \frac{dW}{dt} \qquad (5.1b)$$

$$\frac{dW_B}{dt} = \eta_B \frac{dW}{dt} - \gamma_B \eta_B \frac{dW}{dt} \qquad (5.1c)$$

where η_i are the partitioning coefficients and γ_r and γ_B denote the fraction of imported assimilate respired by roots and stems plus branches (bole + branches). The calculations partition *net* photosynthesis, so there is no explicit respiration.

McMurtrie and Wolf (1983) have presented solutions and analysis of the implications of equations (5.1); for present purposes we will use the equations in finite difference form, summing over a year so that ΔW = total annual dry matter production. We have therefore

$$\Delta W_f = \eta_f \Delta W$$

$$\Delta W_r = \eta_r \Delta W(1 - \gamma_r)$$

$$\Delta W_B = \eta_B \Delta W(1 - \gamma_B)$$

where $\Delta W = 3.65 \times 10^4$ kg ha^{-1} yr^{-1}.

Reasonable values of η_f, η_r and η_B are 0.2, 0.2 and 0.6 (see Figs 2.5 and 5.4) respectively. To estimate values for γ_r and γ_B is more difficult (respiration is discussed in some detail in the next section) but they may both be set to 0.3 (see Jarvis and Leverenz, 1983). Using these values we have

$$\Delta W_f = 0.2 \Delta W = 7300 \text{ kg ha}^{-1} \text{ yr}^{-1}$$

$$\Delta W_r = 5110 \text{ kg ha}^{-1} \text{ yr}^{-1}$$

and

$$\Delta W_B = 15\,330 \text{ kg ha}^{-1} \text{ yr}^{-1}.$$

If we take a value of 0.286 kg m^{-2} for leaf weight per unit leaf area, W_f is equivalent to an increase in L^* of 2.55. This takes no account of leaf fall; the real change in L^* would include this term (i.e. $\Delta W_f = \eta_f \Delta W -$ leaf fall). The root growth term (ΔW_r) could be divided into structural root growth and fine root turnover. This is discussed in more detail later in this chapter.

Biomass data suggest that partitioning ΔW_B into main stem and branches in the ratio 0.8 to 0.2 is representative (see later), hence we have the commercial product from the model: wood production $\approx 0.8 \Delta W_B =$ 13 797 kg ha^{-1} yr^{-1}. Taking a wood density value of 500 kg m^{-3} this represents an annual increment of about 28 m^{-3} ha^{-1} yr^{-1}—a good production rate.

The calculation above serves to illustrate the kind of information which can be obtained from a model of this type, the relationships which can be explored by using it and the parameter values required for it.

5 The Carbon Balance of Trees

For example, I commented at the end of the previous chapter on the variation in the values of the parameters of equation (4.10). The consequences of this variation, for stand productivity, can be readily assessed by varying the values of the parameters. Similarly, the consequences of, say, fertilizer applications which may change partitioning ratios to increase foliage area, or of foliar pests or diseases which reduce foliage area or of periods of water shortage which reduce, say, g_s—and hence C_i—can also be assessed.

Such calculations, if intended to provide information about a real forest, would be done using the appropriate values of the parameters for that forest. Before such models can be safely used in this way they must be intensively tested. This is discussed in the final chapter.

5.2 RESPIRATION

Respiration is the process by which organic substances are oxidized to CO_2 and water with the production of ATP and reducing power (NAPH). Respiration rates can be measured in terms of CO_2 efflux or O_2 uptake. It has become common, in recent years, to consider respiration as consisting of two separate components, one associated with growth and the other with maintenance of tissue. The growth component is considered to involve the use of energy to synthesize new tissue while the maintenance component is associated with protein turnover. This view of respiration was first expressed by McCree (1970) in the form of the equation

$$R_d = a_R \, dW/dt + b_R W. \quad (5.2)$$

The constants a_R and b_R have been empirically determined for a number of herbaceous plants (McCree, 1974); a_R is temperature independent while b_R is temperature dependent.

Equation (5.2) has not been applied to trees. In herbaceous plants the maintenance component—and hence the value of b_R—has been determined by monitoring CO_2 efflux in the dark until respiration rates stabilized, but Butler and Landsberg (1981) measured the CO_2 efflux from apple trees in the dark for up to 36 h and did not observe any sign of a fall and stabilization in the rate of efflux. They said: "We were not able to separate growth and maintenance respiration in these large woody plants but were able to establish excellent relationships between respiration rate and temperature". The relationship between respiration rates and temperature was exponential and could be described by

$$R_d = c_R \exp(k_R T). \quad (5.3)$$

Respiration was expressed on a unit surface area basis (see later comment). Another means of expressing the rates of biological processes in relation to temperature is the so-called Q_{10}. This is defined as the ratio between the rate of a process at T (R_{d1}) and at $T+10°C$ (R_{d2}), i.e. from equation (6.3) we get

$$Q_{10} = R_{d2}/R_{d1}$$
$$= c_R \exp k_R (T+10)/c_R \exp(k_R T)$$
$$= \exp 10 k_R.$$

$Q_{10} = 2$ is often cited as a general value. Butler and Landsberg found $k_R = 0.084$ giving $Q_{10} = 2.3$. This is consistent with many values of the Q_{10} for respiration, which generally seems to be more than 2.

A number of measurements of the stem respiration of forest trees have been made. Rook and Corson (1978) measured it on a *Pinus radiata* tree in a controlled environment room, but the most comprehensive single study to date has probably been that of Linder and Troeng (1981), who made continuous measurements of stem and coarse root respiration of a 20-year-old Scots pine in Sweden from January to November. Hourly respiration rates were exponentially related to temperature with a Q_{10} close to 2 (i.e. $k_R = 0.07$; equation (5.3)), but when daily respiration was plotted against mean daily temperature ($T > 0°C$) the relationship was linear. (I assume this is because when measurements are made over periods long enough to average significant changes in temperature, the exponential relationship is masked by fluctuations and, on average, R_d appears to be linearly related to temperature. In effect, respiration is a process with a response time much shorter than a day; cf. Chapter 2.)

Linder and Troeng (1981) found a pronounced variation in the rate of respiration at a given temperature at different times of the season. They attributed this to acclimation of the respiring biomass to temperature and to varying amounts of respiring tissues. Butler and Landsberg (1981) also identified marked seasonal changes in respiration rate. These were apparent as a change in the value of the intercept (c_R in equation (5.3)) but not in the temperature coefficient k_R. Since high values of R_d coincided with periods when physiological activity was high, the seasonal changes can be explained as changes in the rate at which stored material is mobilized in response to the plant's requirements for material for growth and maintenance. It could also be argued that the constant temperature coefficient reflects the temperature dependence of biosynthetic processes in general, as changes in the value of c_R reflect changes in those processes associated with growth.

Havranek (1981) studied seasonal trends in stem respiration of *Pinus cembra* near Innsbruck, Austria. The pattern of his results is similar to that of Linder and Troeng (1981): there was an exponential relationship with temperature giving Q_{10} values which changed with season and ranged from 1.8 to 2.8. However, rates of respiration were higher by a factor of 10 than those reported by others and must be regarded with suspicion.

Because of the largely peripheral distribution of living tissues and the mass of internal non-living tissue in trees, it is essential to express the respiration rate of trees in terms of surface area. The rate per unit mass of tissue will decline in inverse proportion to the radius as the size of stems, branches and large roots increases. This causes a considerable problem in estimating the surface area of the trees in forests. The problem was somewhat alleviated by the work of Whittaker and Woodwell (1967), who made a great many measurements of the surface area of woody plants and forest communities and provided regression equations for temperate forest species, which allow the surface area to be estimated from linear dimensions. Linder and Troeng (1981) give peak respiration rates of 0.07 mg m^{-2} s^{-1} for coarse roots, and 0.05 mg m^{-2} s^{-1} for the stems of Scots pine at 10°C. These rates are much higher than those given by equation (5.3) with $k_R = 0.084$ and $c_R = 0.006$ (see Butler and Landsberg, 1981) from which $R(T=10) = 0.013$ mg m^{-2} s^{-1}.

It is clear that our knowledge of respiration of trees is incomplete and inadequate, so this component will remain an area of uncertainty in simulations of the carbon balance of trees. When collated, the data on seasonal carbon balance of Linder and Troeng (1980), Linder and Axelsson (1982) and Benecke and Nordmeyer (1982) indicate that total respiration losses from branches, stems and roots appear to range from about 25 to 50% of assimilated carbon. Of this a large part (up to half) is generally attributed to fine root turnover, increasingly being recognized as a factor of major importance in the carbon balance of trees. There are two components of loss by respiration in fine root turnover. When roots die their carbon is lost to the tree and is ultimately respired away by micro-organisms. The growth of new roots requires carbon for "building blocks" and for growth respiration. Functional roots, neither growing nor dying, require carbon for maintenance respiration.

All the data available on tree respiration are for conifers in cold or cold temperate regions. There appear to be no data on respiration losses in tropical or sub-tropical forests.

5.3 DRY MATTER PARTITIONING

Equations (5.1) allocates dry matter between the component parts of trees by means of empirical partitioning coefficients. In this section we consider the relationships between plant parts in general terms, and the allometric equations from which the partitioning coefficients can be obtained. Data from the literature are analysed to provide some values of these coefficients. Dry matter consumption by roots is discussed in the last section of this chapter.

The size of plant parts at any time, relative to one another, depends on the previous growth rates of the parts. The growth rate at any time must depend on the rate of supply of assimilates to the organs. Hence for organ i

$$\frac{dW_i}{dt} = Q_i Y_g - R_i \tag{5.4}$$

where Q_i denotes rate of supply of substrate (labile carbohydrate) to organ i, Y_g is the efficiency with which substrate is converted to plant material (dry weight) and R_i is the respiration rate in equivalent units. Integration leads to a value for the weight of organ i at time t.

Equation (5.4) is descriptive but of no assistance in attempts to predict the relative size of organs. The major question is "what governs Q_i?"—i.e. what determines the rate of assimilate supply to a particular organ? The best answer currently available is that the "sink strength"—demand for assimilate or mobilizing ability of an organ—is determined by the physiological activity of the organ at any particular time. This implies that mobilizing ability will vary with the state of plants (whether they are in a reproductive phase, at bud break in spring, etc) and with environmental conditions. It is also clear that the mobilizing ability of an organ must be considered in relation to that of other organs at the same time and in relation to the supply of assimilate available. If the rate at which assimilate is available is Q_T and the sink strengths of foliage, stem and root systems are given by their mass W_i and specific activity μ_{ai}, then

$$\frac{dW_i}{dt} = Q_T \mu_{ai} Y_{gi} - R_i. \tag{5.5}$$

Specific activity is equivalent to the partitioning coefficients (η_i) in equation (5.1) and equation (5.5) therefore says that η_i are dependent on organ size and activity (see also Landsberg, 1981a); it is really a description of what happens, rather than a description of a mechanism. (Note that the conversion factor (Y_{gi}) may be specific to the organ concerned.)

It is difficult to avoid an element of circularity in this argument because the definition of specific activity is, in effect, the rate of consumption of carbon per unit mass of tissue. If the rate of assimilate supply Q_T changes, the pattern of assimilate allocation will remain unchanged if the ratios of the specific activities of the component parts do not change (i.e. if their relative sink strengths remain constant). However, the ratios between μ_{ai} may vary with rates of supply—there may be priorities for different organs—with stage of growth and factors such as temperature. Perhaps the important question is "can μ_{ai} be altered by breeding (experience with cereals suggests it can) or by silvicultural treatments?" If so then the size ratios of different organs could be changed.

For trees we would expect the response time of assimilate partitioning to be of the order days to weeks, so that we should be able to detect these responses in $W_i(t)$ by measuring the relative sizes of organs over such intervals.

One of the most informative studies on sink activity in trees is that by Gordon and Larson (1968), who studied the seasonal course of photosynthesis, respiration and distribution of radioactive carbon (^{14}C) in young *Pinus resinosa* trees. They found that the photosynthetic rates of "old" needles reached a maximum during the period when shoots were extending rapidly and new needles developing. At this time translocation was from the "old" to the new needles and to the shoots. Photosynthesis in new needles increased to a maximum at about the time they finished growing. Shortly before the end of growth import of assimilates from old needles dropped sharply and the new needles also began exporting large quantities of carbon to other plant parts. Translocation to roots was high when buds were just beginning to elongate but decreased rapidly as the mobilizing activity of the new needles and stems increased. After the new needles had become net exporters, translocation to the roots increased. In general, Gordon and Larson's study demonstrates the point made above, namely that the movement of assimilate to different organs of plants at any time depends on the activity of those organs at that time. In their study the trees were grown in glasshouses under reasonably stable conditions; in the field activity may be influenced by factors such as temperature and water status of tissue as well as stage of development.

In most research on the partitioning of dry matter, experimenters harvest the plants and study the ratios between the weights of their component parts. It is obvious that the ratios observed at any time are the end result (integral) of rates of translocation of assimilate to the various component parts. These may vary considerably in time.

Barnes (1979) has elaborated on the approach outlined earlier, which led to equation (5.5), to the point where he could carry out the integration in terms of time. He derived an equation relating the weights of plant parts at any time t to one another. This can be written

$$\ln W_i = a_\eta + b_\eta \ln W_j - c_\eta t. \quad (5.6)$$

The coefficients a_η and b_η relate to initial weights of W_i and W_j and are obtained empirically; c_η—also obtained empirically—incorporates respiration and tissue death rates. Equation (5.6) implies that, for any particular plant at different times in its growth cycle, the relationship between $\ln W_i$ and $\ln W_j$ is not linear, so that the standard allometric equation (equation (3.2)) would not be expected to hold with the same value of the index throughout the life of a tree. However, Barnes points out that a special case of linearity arises when growth is close to exponential (the normal condition with seedlings and young plants). For example, if $W_i = W_i(0) \exp k_i (t - t_o)$ and $W_j = W_j(0) \exp k_j (t - t_o)$, the term involving t may be eliminated leading to

$$\ln W_i = \ln W_i(0) + k_i/k_j [\ln W_j - \ln W_j(0)]. \quad (5.7)$$

This equation was also derived by Landsberg (1981a) starting from equation (5.5). Equation (5.7) may be written in the form of (5.6), which then becomes the usual allometric equation, which is

$$\ln W_i = a_\eta + b_\eta \ln W_j$$

normally written as the power relation

$$W_i = a W_j^n. \quad (5.8)$$

It follows from equation (5.8) that if W_j is the weight of the stem and the relationship between stem mass and diameter at breast height (d_B) is linear, then equation (5.8) is equivalent to equation (3.2) and we can expect to find that

$$W_i = a' d_B^{n'} \quad (5.9)$$

where a' will not have the same values as a in equation (5.8), but $n = n'$. (Forrest (1969) has presented detailed data on the weights and sizes of *P. radiata* of various ages and has shown that the assumption that stem mass is linearly related to diameter at breast height is tenable.)

If equation (5.8) holds throughout the life of trees, it provides the partitioning coefficients (η_i) immediately (by differentiation). In the discussion on stand structure in Chapter 3, we noted that there have been many studies on tree biomass and the relationship between tree parts, these relationships being commonly expressed in terms of equation (5.8). White's

5 The Carbon Balance of Trees

(1981) survey of such studies indicated that the value of n is about 2.5 in most cases. In the following discussion I have analysed some data from the literature to provide detailed information on allometric ratios and the application of some of the equations presented above.

An interesting study of biomass amounts and distribution was made by Fujimori *et al.* (1976); some of their data have been reprocessed and are shown in Fig. 5.3, where branch and foliage weights of a number of species

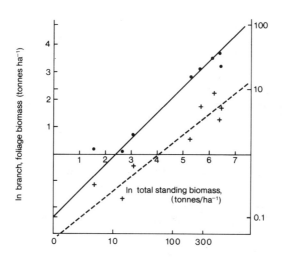

Fig. 5.3. Branch (●) and foliage (+) biomass of several species (*Tsuga heterophylla, Picea sitchensis, Abies amabilis, Acer macrophyllum*) in the northwestern USA, plotted, on logarithmic scales, against total standing biomass. (Derived from data tabulated by Fujimori *et al.*, 1976.) (Back-transformed values are given for reference.) The lines were fitted by eye. The slope of the branch mass/total mass line is about 0.97, that of the foliage mass/total mass line is 0.83. In view of the greatly varying ages of the trees, and the different species, the small scatter of the data is remarkable.

have been plotted on logarithmic scales against stem weights. Since these data are from different stands (not growing exponentially) and different species the relationships—especially the branch weight/stem weight relationship—

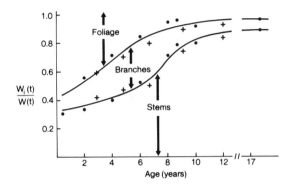

Fig. 5.4. Normalized plot of the time course of the ratio of the dry mass of foliage, branches and stems ($W_f(t)$, $W_b(t)$, $W_B(t)$) to total above ground mass ($W(t)$) for *Pinus radiata*. The data are from age series studied by Madgwick *et al.* (1977) in New Zealand (●) and Forrest (1969) in the ACT, Australia (+). The curves were drawn by eye. (The proportion of foliage at any time is $(1-(W_b+W_B))/W$... etc.) Distances between curves indicate the proportion of each component. The data indicate that these ratios change with age, up to about full canopy.

are remarkably good; certainly the data provide no reason to reject equation (5.8) in favour of (5.6). The relationship between branch weight (W_b) and stem weight (W_B) is nearly linear ($W_b \propto W_B^{0.97}$, i.e. $W_b \approx$ constant $\times W_B$). The constant was approximately 0.1 so we can take it that for these stands $W_b \approx 0.1 W_B$.

Time series data for trees have been published by Ovington (1957), in a thesis by Forrest (1969), by Madgwick *et al.* (1977) and by Albrektson (1980). Forrest harvested 3-, 5-, 7-, 9- and 12-year-old *P. radiata* in the Australian Capital Territory and Madgwick *et al.* harvested *P. radiata* trees in nine age classes between 2 and 22 years old in New Zealand. From the data of these authors it is possible to obtain average branch, stem and foliage weight per tree at time of harvest. Figure 5.4 shows these as a fraction of total dry mass at time t, i.e. $W_i(t)/W(t)$. The data of Madgwick *et al.* suggest that the ratios change with age. Data obtained by Ovington (1957) for *Pinus sylvestris* (not shown) indicated a gradual decline in W_f/W until about age 20–25 years; there was then a large, abrupt fall, after which W_f/W remained stable at about 0.05. The other notable change was a steady increase in W_B/W up to about 0.55–0.6 at age 30, after which it remained stable. For the Forrest (1969) and Madgwick *et al.* (1977) data, the weights of branches and foliage were plotted against stem weight on logarithmic scales (not shown). There was some evidence of a discontinuity in the New Zealand data at about age

8–10 years, but the Australian data fall on a straight line except for the point from the 7-year-old trees. The slopes (n) of lines drawn through the branch weight/stem weight points for both sets of data were about the same, although the intercepts were different. The relationships were very approximate because of the paucity of data points, but gave $W_b \approx$ constant $W_B^{0.66}$ in both cases. In the units used (tonnes ha^{-1}) the intercept has a value of about unity in the New Zealand data but was less in the Australian data.

A good set of time series biomass data—including coarse root weights—is provided by Albrektson (1980), who harvested trees from 15 *Pinus sylvestris* stands in Sweden, ranging in age from 7 to 100 years. Root and branch weights were plotted against stem weight on logarithmic scales, which showed that the data are consistent with equation (5.8), with index n about 1.0 and 0.6 respectively. The intercept of the root weight/stem weight line is about the logarithm of 0.1, hence (as in the case of the branch weight/stem weight relationship derived for different species from the data of Fujimori *et al.*) $W_r \approx 0.1 W_B$. The similarity in the branch weight/stem weight index in the *Pinus sylvestris* and *P. radiata* data, as well as the species studied by Fujimori *et al.* (1976), suggest that the numerical values may be generally applicable. This is at least worth further investigation.

In Fig. 5.5 root, branch and stem weights from Albrektson's *Pinus sylvestris* data have been plotted on logarithmic scales against total tree weight ($W(t)$). The lines drawn through the stem weight/total weight and root weight/total weight points both have slopes of about 1.2. The branch weight/total weight relationship departs from linearity when total weights are small—the younger stands. This indicates, as do the data of Madgwick *et al.* (1977), that allometric ratios may change with age and equation (5.6) may have to be used for accurate prediction.

If allometric ratios are given by equation (5.8) then the partitioning coefficients are given by $dW_i/dW = a n_i W^{n_i-1}$. Therefore if $n \approx 1$ the arithmetic ratios of the weights of component parts will not vary with W; if $n \neq 1$ then the ratios vary. In the case of the *Pinus sylvestris* data (Fig. 5.5) $n \approx 1$ for the coarse roots and stems, implying that these components comprise more or less constant proportions of plant weight as the tree grows. This was also the case in Ovington's (1957) *Pinus sylvestris* data.

The relatively small amount of empirical data assembled here generally supports the use of equation (5.8), with constant parameter values for describing the relationships between the component parts of trees. However, there are indications that this may not always be adequate and further, more precise investigations are needed. Ideally changes in dry matter partitioning should be investigated in trees of varying ages similarly treated, and in trees of the same age grown at markedly different spacings and also subjected to different fertilizer and water treatments. In Chapter 3 I discussed the

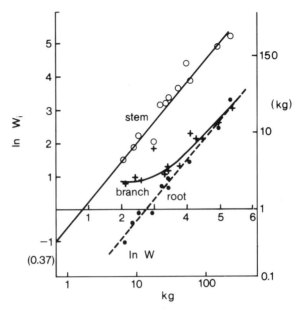

Fig. 5.5. Root (●), stem (○), and branch (+) weights, from an age series of *Pinus sylvestris*, plotted on logarithmic scales against total weight (kg) per tree. (Back-transformed values are given, for reference, on the right-hand side and at the bottom.) The data are from an age series (9- to 100-year-old stands) studied by Albrektson (1980). The lines are drawn by eye; the two for stems and roots are parallel, with slopes of about 1.2. The curve in the line showing the branch weight/total weight relationship indicates the greater mass of branches needed to support large foliage masses on young trees.

dependence of total biomass (per tree) on population; it is likely that competition will affect not only the weight of trees but also the relative weights of their component parts. This is supported by Pearson *et al.* (1984), who derived the allometric equations for estimating leaf area and bole, branch, foliage and root biomass of *Pinus cortonta* trees grown at different stand densities. They conclude: "clearly the biomass distribution of lodge-pole pine trees varies with forest structure and site characteristics". Note that equation (5.7) provides a useful basis for interpreting the significance of a_η and b_η.

The relationships described by allometric equations are static, describing the state of the tree or stand at a particular time. In reality such relationships are never truly static; there is growth of various components and losses of above-ground parts (litter fall). On average, annual litter fall in temperate forests is made up of about 70% leaves and 7% each of branches, bark and fruits (Jarvis and Leverenz, 1983). Variation between species and localities is large.

5 The Carbon Balance of Trees

The coarse root weights given by Albrektson (1980) have been plotted, on logarithmic scales, against total tree mass in Fig. 5.5. The data indicate that root weights vary only slowly in relation to total biomass. Pearson *et al.* (1984) established linear regressions between root mass and tree basal area. Fine root biomass, which may constitute only a small — but very important — proportion of total root biomass, may vary widely in time; the time scales of growth and regeneration of fine roots are apparently much more rapid than those of any other part of the plant. Variations in root system mass are discussed in the following section.

5.4 ROOT MASS AND TURNOVER

Root systems support the plant and provide the means for absorbing water and nutrients from the soil. Analyses of their effectiveness in the performance of these functions cannot therefore be made in terms of root mass only but must take into account the structure of the root system. For adequate support, trees need root systems that spread wide enough and penetrate deeply enough to hold the tree in place in all normal situations. The structure which will meet these requirements varies with soil type and condition. Circumstances may arise (e.g. wet soil, high winds) where the support provided by roots is inadequate and trees blow over, but I am not here concerned to analyse such situations.

To be efficient absorbers of water and nutrients, roots need large absorbing surfaces, therefore we might expect that the assimilate channelled to root systems would always be used to produce thin roots with high ratios of surface area to mass. Plants adapted to relatively wet fertile soils may well have evolved such root systems, which would give them a competitive advantage over plants that produced fewer, larger-diameter roots. However, if the soil in the root zone becomes dry, a system of fine fibrous roots with large surface area would be a disadvantage. Such roots, in which there is high resistance to water flow, have to be relatively short, otherwise they cannot carry enough water to the stem. Therefore the volume of soil explored would be relatively small, although it would be thoroughly explored. (Fine, fibrous root systems are characteristic of grasses. Most grasses are annuals and their root systems enable them to exploit very effectively the water and nutrients available during their favourable growing periods.)

Water movement along roots is governed by potential gradients along the roots and the resistances in the flow pathways; these are (approximately) inversely proportional to the reciprocal of the fourth power of the xylem element radius (Poiseuille's law). For a given mass of assimilate available for root formation, there is an optimum root-length/root-diameter relationship.

This has been analysed by Landsberg and Fowkes (1978), who were able to provide a mathematical definition of this optimum for a single root with laterals. They later (Fowkes and Landsberg, 1981) extended their analysis to examine root systems as a whole, and tried to determine whether they conformed to some "optimal" criteria. It emerged from the analysis (which was made in terms of water movement, not nutrient uptake) that the best root system for a plant in a given environment—and therefore, presumably, the one most likely to have evolved in that environment—will depend on the transpiration rates likely to occur there and on the water potential that the plant can tolerate (see Chapter 7). It will also depend on soil type—the hydraulic conductivities of light and heavy soils when dry and wet are vastly different.

There has been some controversy about whether the form and distribution of root systems is genetically determined or controlled by soil conditions. It seems clear that both factors can influence the results. Grass plants do not produce tap roots or a few large thick roots, and most trees do not produce root systems consisting only of fibrous roots—in other words we expect the general form of root systems to be characteristic of species and hence genetically controlled. However, the size and distribution of a root system is certainly affected by environmental conditions such as water and nutrition. The roots of most forest trees will not normally grow in water-logged soil and they will not grow in dry soil. Roots tend to proliferate in regions of high fertility and suitable moisture status.

Root systems are not static; trees do not simply produce a root system of a particular type and conformation and sustain it under all conditions. There is constant adjustment: when soil conditions are non-limiting, assimilate is diverted to aerial growth; if the rate of uptake of water and/or nutrients is inadequate to sustain the potential growth rate of the tree, then assimilates will be diverted to increase the absorbing root surface areas, provided this is possible in the prevailing soil conditions. (This implies that the specific activity of roots varies—see earlier discussions.) We know very little about the signals which elicit changes in the pattern of assimilate allocation to root (e.g. changes in growth regulators) but the partitioning model presented earlier (equation (5.5)) is, in principle, satisfactory. If there are no limitations to foliage production, cambial or meristematic activity, and no nutrient shortage, then the specific activity of aerial parts will be greater than that of roots and top growth will be favoured. The reverse also applies.

It follows from these arguments that the study of root systems must involve the study of their dynamics. This is necessarily tedious, time consuming and difficult. However, the amount of good data in the literature on tree root systems has increased. Some of it will be discussed here to illustrate the points made and to provide some information on the amounts

5 The Carbon Balance of Trees

of assimilate involved in root production and turnover. The experimental methods used to study roots and root systems are discussed in the papers cited and also by Böhm (1979).

There are many estimates of the root mass of trees at any particular time in their growth cycle. One of the most comprehensive studies is that by Jackson and Chittenden (1981), who completely excavated the root systems of a number of P. radiata trees that were from 3 to 8 years old. They separated the roots into very large ($W_{r.1}$; > 20 mm diameter), large ($W_{r.2}$; 10–20 mm diameter), medium ($W_{r.3}$; 5–10 mm diameter), small ($W_{r.4}$; 2–5 mm diameter) and fine ($W_{r.5}$; < 2 mm diameter). This classification is identical to that advocated by Böhm (1979), except that he had a further category of "very fine roots" (< 0.5 mm diameter). Such consistency between classification systems is important to ensure comparability of results.

Jackson and Chittenden (1981) produced a series of regression equations from which the weight of roots in each size class could be calculated from a measurement of another part of the tree. A number of such independent variables were used. In the case of fine roots—arguably the most important from the point of view of water and nutrient absorption—the best relationships were obtained between fine root dry weight ($W_{r.5}$) and foliage weight or the weight of "small" roots ($W_{r.4}$). The equations were linear in both cases. They are given here with all (dry) weights in kilograms;

$$W_{r.5} = 0.08 + 3.6 \times 10^{-3} W_{r.4} \tag{5.10}$$

$$W_{r.5} = -0.12 + 5.3 \times 10^{-3} W_{f} \tag{5.11}$$

If the total mass of a root system varies over a relatively short time interval it will be the fine root component that varies most. Because of their small size, fine roots are more likely to be transient than larger (support) roots. Relationships such as equations (5.10) and (5.11) should therefore be expected because the "small" roots provide the immediate support structure for fine roots and foliage provides the assimilate necessary to produce them.

The following equations derived by Jackson and Chittenden (1981) give the mass of roots (per tree) greater than 2 mm in diameter ($W_{r.4}$), less than 5 mm ($W_{r.3}$), and total root mass (W_r), in terms of stem diameter d_B:

$$W_{r.4} = 6.25 \times 10^{-3} d_B^{2.74} \tag{5.12}$$

$$W_{r.3} = 5.97 \times 10^{-3} d_B^{2.81} \tag{5.13}$$

$$W_r = 5.87 \times 10^{-3} d_B^{2.94}. \tag{5.14}$$

For each size class, root length can be estimated from root dry weight; Jackson and Chittenden found that the standard allometric equation again provided the best results. Writing this as $L_i = a_r W_{r,l}^{m_r}$ the values they gave for a_r and m_r are

Root class	large	medium	small	fine
a_r (×10³) (kg)	1.68	7.7	58.2	851.2
m_r	0.991	0.958	0.820	0.571

To analyse the effectiveness of root systems, investigators often use parameters such as root length per unit soil volume (L_V) or root length per unit soil surface area (L_A) (analogous to L^*). Conversion factors such as those tabulated above are therefore essential.

Jackson and Chittenden's (1981) study provides invaluable data but no information on root dynamics. This has come in recent years from work by Ford and Deans (1977), Deans (1979, 1981), Persson (1978, 1980a,b) and others.

Ford and Deans found that there was no net increase in the weight of fine roots (in this case less than 0.5 mm in diameter) of Sitka spruce over a growing season, although the weight of fine roots varied from 20 to 34 kg m^{-3} (L_V varied from 18×10^3 to 32×10^3 m m^{-3}). Root carbohydrates increased in early summer in a manner which suggested that the build-up was a result of export from the previous season's shoots. The minimum soluble carbohydrate concentrations in the storage roots were reached one week after fine root weight reached its maximum, suggesting that fine roots grew at the expense of labile carbohydrates.

Some of Deans' (1979) data, illustrating the effects of temperature and soil moisture on fine root growth in Sitka spruce in Scotland, are shown in Fig. 5.6. He noted that root mortality occurred whenever the water potential of the soil fell significantly below about -0.01 MPa. In a later study on the dynamics of coarse root production, Deans (1981) found that the youngest and thinnest roots (in size classes greater than 5 mm diameter) grew fastest and also contributed most to root biomass. Collating data from several studies Deans concluded that, for the 10-year-old spruce trees studied by him and by Ford and Deans (1977), fine root production was 1.38 kg tree^{-1} yr^{-1}. Production of intermediate roots (2–5 mm diameter) was 0.01 kg tree^{-1} yr^{-1} and of thick roots (greater than 5 mm diameter) was 0.83 kg tree^{-1} yr^{-1}. On a land area basis, these values indicated that total root production exceeded 8 tonne ha^{-1} yr^{-1}, which was about 34% of mean root biomass. Deans did not give data on total biomass production but if we assume that it was roughly the same as the annual production by *Pinus sylvestris* of similar age in Sweden—i.e. about 11 tonne ha^{-1} yr^{-1} in an untreated plot (Linder and Axelsson, 1982)—then total root production constituted about 70% of

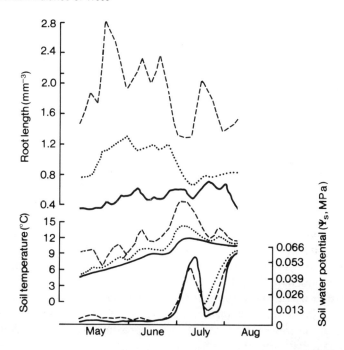

Fig. 5.6. The upper three lines show the variations in the quantities of fine roots of Sitka spruce (*Picea sitchensis*) observed in the top (needle/litter layer, dashed lines), the turf layer (under the planting ridge, dotted lines) and the top 0.10 m respectively of an undisturbed peat horizon (solid lines) by Deans (1979). The lower two sets of lines show the variations in soil temperature and soil water potential in the same layers during the same period. Deans found the main environmental effect to be the significant correlation between fine root biomass and soil water potential. (Note that the ψ_s values are plotted as positive.)

annual production and fine root production used 45% of the assimilates available for growth.

Grier *et al.* (1981) measured biomass distribution and above- and below-ground production in young (23-year-old) and mature (180-year-old) *Abies amabilis*-dominated ecosystems in the Cascade Mountains in Washington State, USA. Net primary (biomass) production by the ecosystems was 18 tonne ha^{-1} yr^{-1} in the young stands, of which 65% was below ground. Production by the mature stand was 16.8 tonne ha^{-1} yr^{-1} of which 73% was below ground. Conifer fine root production was 36% of primary production in the young and 66% in the mature stand. Grier *et al.* found that fine root biomass in *Abies amabilis* reached peak values in early spring and late autumn with low values in summer and winter. The increases in fine root biomass in the spring occurred in both the young and mature stands well before any sign of above-ground growth activity, and root growth in general appeared to be independent of aerial growth. These data again illustrate the principle of allocation in the direction of specific activity; activity in the roots commences in the spring and they become the primary sink for carbohydrates. As shoot growth commences, the total activity in the shoots becomes greater than in the roots and available carbohydrates move towards the shoots. At other times in the growing season the direction of carbohydrate flow (whether to roots or to tops) depends on the relative activity of the plant parts.

Keyes and Grier (1981) tested the hypothesis that trees on sites short of water or nutrients expend comparatively more energy on below-ground dry matter production than trees on more productive sites. They studied 40-year-old Douglas fir stands on high and low productivity sites, where the standing stem biomass was 423 and 231 tonne ha^{-1} respectively. Above-ground productivity was estimated from direct increments of tree growth and below-ground productivity was derived from a combination of core sampling and observation windows. Keyes and Grier did not find statistically significant differences in the mass of small and fine roots on the two sites, but fine root turnover at the low productivity site was approximately 2.5 times that on the high productivity site.

Their study supports the suggestion that a greater proportion of the available assimilate will be diverted to fine root growth where soil conditions are limiting for growth. In the study by Keyes and Grier the annual above-ground and coarse root biomass increment on the poor site was approximately half (6.4 tonne ha^{-1}) that on the good site (12 tonne ha^{-1}) but primary productivity differed by only 13% when all above-ground and below-ground components of the two stands were taken into account (15 tonne ha^{-1} yr^{-1} on the low productivity site; 18 tonne ha^{-1} yr^{-1} on the high productivity site). However, 36% of the dry matter on the low productivity site went into fine root turnover with 13% going to foliage. The corresponding figures for the high productivity site were 8 and 3.4%. The differences in stem biomass at the two sites can therefore be attributed to the integrated differences in dry matter partitioning over the life of the two stands.

The results of Keyes and Grier (1981) are supported by those of Persson (1980b), who studied fine root dynamics in Scots pine stands with and without near-optimum nutrient and water regimes. His data provide further evidence for the sensitivity of fine roots to water stress. He found that in the irrigated and fertilized plots fine root production was slightly greater than in the control plots, while above-ground production in the irrigated and fertilized plots was nearly double that in the control plots. Therefore, as a proportion of total productivity, fine root turnover was much greater on the control plots. Linder and Axelsson (1982) estimated from Persson's data that fine root production in the control plots used about 35% of net assimilate as opposed to about 11% in the irrigated and fertilized plots.

All these studies indicate the tremendous importance of fine root turnover in the carbon balance of trees and it is clear that any productivity estimates that ignore roots may be grossly in error. It is also clear that the rapid turnover of fine roots provides a very flexible mechanism for responding to changing conditions. If, for example, trees are on a site which is drying, the fine roots may die off in the upper layers, since the soil will tend to dry from the top downwards (partly as a result of high root densities in those layers), but they may proliferate in lower layers as the damp zone retreats. The amount of structural root available in lower layers is less than in the upper layers, so that the potential for fine root production is smaller in lower layers. With less fine roots, the tree's ability to take up water is decreased, consequently the resistance to transfer of water from soil to roots is increased and water stress increases. The questions of root resistance to water uptake and the importance of water potential in these matters are discussed in the chapter on water relations (Chapter 7).

5.5 CONCLUDING REMARKS

The earlier parts of this chapter brought together some of the concepts outlined in Chapters 3 and 4. Radiation interception and equations for leaf photosynthesis were used to calculate canopy photosynthesis, and hence dry matter production. This brief excursion into production modelling served to illustrate how this kind of knowledge can be used to explore the consequences of changes in canopy structure and the effectiveness with which leaves convert radiant energy into carbohydrates. To estimate net production, losses of carbon as a result of respiration have to be considered. The information reviewed suggests that much more work is required in this area. Losses due to shedding organs (leaves, branches) and mortality of individuals, have not been considered, but for calculations covering periods of years these too would be included in the accounting (see Jarvis and Leverenz, 1983).

Information about the total productivity of a stand is only of value if we know how much of the product is useful, so consideration of assimilate (dry matter) partitioning follows logically from the discussion of stand productivity. We are currently far from a useful explanation of the mechanisms controlling assimilate partitioning but ideas about the specific activity of organs are useful and new analytical developments (see Lang and Thorpe, 1983) may contribute to the value of empirical data. Meanwhile, there is a great deal of data available on the biomass of forest trees and analysis of these in terms of allocation ratios can provide valuable information.

The concept of harvest index—the ratio of the weight of useful product (stem) to total biomass—is a useful guide to carbon partitioning. It is widely used in agriculture and (implicitly) by tree breeders when they select for small, thin branches. However, there are few good values for trees because the roots are not included in most production studies. Jarvis (1981) obtained harvest index values between about 60 and 70% from published data.

An area which requires much more exploration, in this respect, is variation in allometric ratios with varying tree populations. There has been little attempt to question the consequences, for both rates of production and partitioning of assimilates, of populations very different from the usual range.

The review of root mass and fine root turnover data supports the argument that specific activity of organs changes with changing conditions. Evidence currently available indicates that more assimilate is diverted to roots in poor growing conditions than in good conditions. We need to investigate this in much more detail, and establish functional relationships between rates of root production and turnover and various environmental conditions such as soil wetness and fertility. It may also be necessary to take account of the effects on assimilate partitioning of conditions in the aerial environment, such as air humidity, which reflects the drying power of the air and, through its influence on transpiration rates and hence tree water status (see Chapter 7), may exert a significant effect on the patterns of tree growth.

Another area which is of major importance is the heritability of different assimilate partitioning coefficients (see equation (5.1)). Whether there are large differences amongst genotypes in the way they allocate carbon to different organs needs to be known and exploited by those concerned with tree improvement. We do not know the extent to which harvest index is influenced by environmental conditions or whether it can be changed by selection and breeding.

The discussions in this chapter have been concerned with carbon balance *per se*, without explicit discussion of factors such as nutrition. The influence of nutrition (and tree water relations) is, however, implicit in varying factors such as L^* and the leaf photosynthetic parameters. The increased productivity that results from improved nutrition of trees can, very largely, be attributed to the effects of improved nutrition on leaf growth—and hence leaf area—and photosynthetic efficiency. However, these effects are not well documented, and most studies on nutrition have dealt with empirical responses of forest stands to applied fertilizers.

Nutrient dynamics, and the effects of nutrition on tree growth, form the subject of the next chapter.

6 Nutrient Dynamics and Tree Growth

The nutrient reserves of a forest are held in the foliage, bark and branches, sapwood and heartwood of the trees, in the soil and in the understorey vegetation. A small fraction of these nutrients is in flux between the various compartments of the system. An understanding of the dynamics of nutrients, or nutrient cycling, is an essential prerequisite for understanding and predicting the effects of nutrition on forest growth. Appreciation of this has led to a considerable amount of research, in recent years, on nutrient contents, forest biomass, and nutrient dynamics.

As in any complex biological situation it is possible to consider the system at a number of levels, although the response times of the processes at these levels are not well known. Much nutrient cycling work is based on studies of litter fall and decomposition, in which collections and analyses are made at intervals of weeks or months (e.g. Miller *et al.*, 1976a), or on biomass sampling at intervals of years (Attiwill, 1979; Forrest, 1969). These intervals obviously do not reflect the turnover times of nutrients in all components of the system. Nutrient movement from soil to leaves takes place continuously, but the analytical techniques commonly used cannot identify the short-term changes. Similarly, translocation between organs is continuous but mass changes can only be determined with any accuracy over intervals of days or weeks; for example, Fife and Nambiar (1982) followed changes in nitrogen and phosphorus concentrations in *Pinus radiata* needles over periods of weeks. They found that in spring the initial mass accumulation rate of nitrogen was about seven times that of phosphorus, but the rate of withdrawal of nitrogen in summer was only twice that of phosphorus (0.011 microgramme nutrient per microgramme dry weight per day for phosphorus and 0.021 for nitrogen). However, we note that on a molar basis, which will be a better reflection of chemical activity, the ratio was about 3 in spring and unity in summer (0.35 mol μg^{-1} day^{-1} for P and 0.29 mol μg^{-1} day^{-1} for N).

Leaves remain on trees for periods ranging from one season, in deciduous trees, to 10 or more years in conifers. Litter on the ground may decompose and release most of the nutrients it contains within a few months in a tropical forest, or it may require years in cold regions. Except in very young soils the weathering of soil minerals and the release of nutrients into the soil solution is generally an extremely slow process, requiring many years to add a significant quantity to the available pool. The uptake of nutrients from the soil solution, by plant roots, is a continuous process. The rate at which nutrients are required by trees must depend on the growth rates, determined by radiation interception and the factors considered in Chapters 4 and 5. However, the rate at which nutrients are actually absorbed by trees depends on the accessibility of the nutrients to the root systems, and their chemical availability. This does not imply that the short-term growth rates of trees are directly proportional to nutrient uptake rates; as noted above nutrients within trees may be quite rapidly remobilized and translocated between tissues (see 6.1.3, "biochemical" cycling). Nevertheless, in the longer term, the amount of dry matter produced must be proportional to the mass of nutrient absorbed.

Under steady state conditions the rate of nutrient uptake by plants in stable ecosystems will balance inputs from weathering of minerals and decomposition (oxidation) of organic matter. Steady state conditions are unlikely to occur over short periods (up to seasons) in any system, and total nutrient gains and losses from forests intensively managed for wood production are unlikely to be equal. Management of these systems must therefore have, as one of its main aims, the achievement of a balance between uptake and mineralization processes. When these are out of balance both the chemical and mass equilibria of nutrients in the soil become disturbed, with possibly serious consequences, such as soil acidification and nutrient depletion.

The next section provides a general discussion of nutrient cycling; the remainder of the chapter deals in more detail with various components of the nutrient cycle and then with tree growth in relation to nutrition.

6.1 NUTRIENT CYCLING

A simplified outline of nutrient cycling of a forest is presented in Fig. 6.1.

Switzer and Nelson (1972) proposed that the circulation of nutrients in forests should be defined in terms of three cycles:

(1) *geochemical cycles*, which encompass the gains and losses of nutrients to the ecosystem by processes such as leaching and weathering;
(2) *biogeochemical cycles*, which encompass soil–plant relationships, including nutrient gains to the soil by symbiotic fixation and organic matter decomposition, and losses by uptake by plants;
(3) *"biochemical" cycles*, which encompass internal transfer relationships or

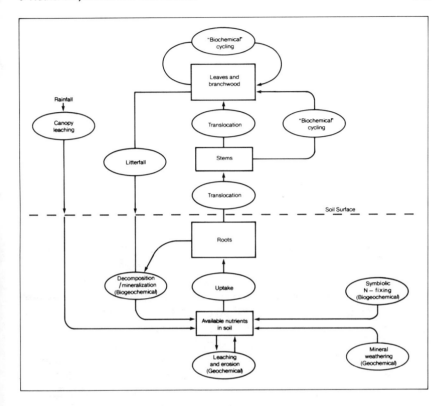

Fig. 6.1. Diagrammatic representation of nutrient cycling.

translocation of nutrients within the vegetation. (This term can cause confusion with the biochemical cycling of organic compounds, hence the quotation marks to distinguish it.)

These cycles are labelled in Fig. 6.1 and the processes involved in them will be discussed in sequence. The geochemical and biogeochemical cycles will be considered in relation to the volume of soil exploited by roots so that addition of nutrients to that volume, or their removal from it, are taken as gains or losses respectively.

6.1.1 The Geochemical Cycle

Geochemical processes are usually slow relative to the growth of trees. Nutrients are added to the soil by the weathering of parent materials, and in rainfall. Charley (1981) collated data on nutrient inputs in precipitation, from a number of Australian sources. Mean values were 3.7 kg K, 6.4 kg Ca, 5.2 kg Mg and 5.3 kg s ha^{-1} yr^{-1}. Baker and Attiwill (1981) give 3 kg ha^{-1} yr^{-1} as an average value for N input, in rain, to eucalyptus forests in Australia, while Miller (1981) gives values between 5 and 22 kg N ha^{-1} yr^{-1} for N inputs by rainfall at sites in Scotland and Europe.

The rate of weathering and the consequent release of nutrients from the soil are quantitatively poorly understood. Pritchett (1979), who provides a useful summary of data for forest soils, notes that the rate of release depends on the nature and the amount of mineral reserves and on environmental conditions such as temperature and wetting and drying cycles. Inputs of N and P from weathering are usually small, especially in old highly weathered soils. Wood et al. (1984), referring to a hardwood forest in the northern USA, state that "more than 10 kg ha^{-1} yr^{-1} of organic phosphorus are mineralized through decomposition processes in the soil". Values for the rates of release of K, Ca and Mg, collated by Miller (1981), are 4–7, 12–21 and about 4 kg ha^{-1} yr^{-1} respectively.

Nitrogen and cations are added to the soil directly in rainfall. They are also recycled. Rainfall intercepted by foliage and branches, which then drips down through the canopy or runs down stems, leaches nutrients from the trees and returns them to the soil. Miller (1984) reviewing data on this process—which is, strictly, biogeochemical, not geochemical—concludes that crown leaching is a significant process but that its importance varies between nutrients. Baker and Attiwill (1981), from various sources, give values around 5 kg ha^{-1} yr^{-1} for nitrogen. Data from Pastor and Bockheim (1984) give a ratio of 2.9 between nutrient concentration in rainfall and throughfall.

The main natural mechanisms of nutrient loss from the soil–plant system are leaching of soluble nutrients out of the root zone and losses in surface erosion. Surface erosion may consist of movement of individual particles or mass flow (down-slope movement of large masses of soil). The latter mechanism will obviously cause much greater nutrient losses. Miller (1984) concluded that leaching losses are generally negligible in most undisturbed forest ecosystems; nutrients are retained by micro-organisms or absorbed by tree roots before they leave the system. However, on sandy soils leaching may be responsible for significant loss of fertilizers.

6.1.2 The Biogeochemical Cycle

The biogeochemical cycle describes the transfer of nutrients between the soil and vegetation. It includes gains of nutrients to the soil by decomposition of organic matter, and losses by uptake by plants, by burning and by wood removal. Only a quarter to a third of the nutrients taken up by trees accumulate in standing biomass over the life of a forest; the remainder are released in litter and root turnover and crown leaching (Miller, 1984). The rate of nutrient addition by decomposition of organic matter depends on the environmental conditions and the amount and type of nutrients released from organic material depend on the mass and composition of the material. Since decomposition is carried out by micro-organisms, the nutrients in organic matter may not be released into the soil solution immediately but may be immobilized by the micro-organisms for a period. If—as in tropical forests—the process of decomposition is rapid and essentially continuous, and the system is in equilibrium, the relatively small amounts of nutrient retained in decomposing organic matter will not be important for tree growth. However, if there has been a major input of organic matter to a system, such as might occur after logging, there may be a period during which significant amounts of nutrients will be extracted from the soil, and unavailable to plants, to provide for a massive increase in microbial populations in response to the input of decomposable carbon. In regions where microbial activity is inhibited by low temperatures or dry periods this period of nutrient unavailability could extend across several growing seasons, becoming a significant factor in re-establishment. Ironically, the final result of logging may be a reduction in soil organic matter, brought about by the stimulation of microbial activity by the higher temperatures and water content of uncovered soil and the better aeration resulting from disturbance.

Nitrogen. In the forms useful to plants nitrogen is probably the nutrient most universally limiting to plant growth. It is also most vulnerable to loss and its biogeochemical cycle warrants particular attention. Most nitrogen in the soil was originally fixed by symbiotic and non-symbiotic organisms, but the rate of supply (addition to the soil) from these sources is usually relatively slow in relation to rates of extraction by trees (Davey and Wollum, 1979) and the major source of the nitrogen in the soil solution at any time is decomposing organic matter. Typical estimates of rates of symbiotic fixation in eucalyptus forests are in the region of 10 kg ha^{-1} yr^{-1} (Baker and Attiwill, 1981). Pastor *et al.* (1984), in a thorough study of forest productivity in relation to N and P cycling, found N-mineralization rates in mixed forests in Wisconsin, USA, from 26–84 kg ha^{-1} yr^{-1}. Soil N-mineralization was positively correlated with litter production and N and P return in litter.

Information reviewed in Chapter 5 indicated that large amounts of carbohydrate are utilized to form fine roots, which may decay after a short time. It follows that there must be a rapid turnover of the nutrients associated with these organs, but there is very little information on the contribution of fine-root decay to soil nutrients. McClaugherty et al. (1982) estimated the nitrogen requirements for fine root production in a 53-year-old stand of *Pinus resinosa* and an 80-year-old mixed hardwood stand and found that these approached the estimates of total N-mineralization and uptake in hardwood forest. They cite measured mineralization rates in temperate forests ranging from 50 to 300 kg ha^{-1} yr^{-1}, with an average of about 100 kg ha^{-1} yr^{-1}, while demand by above-ground vegetation in temperate forests falls in the range 50 to 150 kg ha^{-1} yr^{-1}. McClaugherty et al. therefore conclude that, if the estimates of fine root production are to be accepted, large amounts of nitrogen must be retranslocated from senescing fine roots to above-ground parts of plants. This may well be the case (see discussion on "biochemical" cycling), although these estimates of N-release rates cannot be taken as accurate.

Nutrient uptake. Nutrient uptake by plants is part of the biogeochemical cycle. In general terms, the amount taken up over an interval such as a growing season must be proportional to the dry matter increment. Obviously there are many caveats and qualifications to that statement: for example, the relationship between nutrient uptake rate and growth of a young forest is unlikely to be the same as that of an old, stable ecosystem. In particular retention in tissues, and transfers between tissues and in litterfall, would be very different. Some of these factors will be discussed later, but the proportionality between growth and nutrient uptake is strongly supported by data collated by Miller (1984) (Fig. 6.2) and by the highly significant correlation between net above-ground production by a mixed forest, and rate of N-mineralization, obtained by Pastor et al. (1984). It provides a means of analysing the nutrient uptake component of the biogeochemical cycle. If we can estimate dry matter production, and if we have values for the nutrient concentration in the dry matter, we can estimate uptake by trees. Formally denoting the average mineral nutrient concentrations in trees as [M] (mass minerals/unit mass dry matter) we can write

$$\text{Uptake rate per unit land area} = [M]_p \, dW/dt \qquad (6.1)$$

where p is tree population per unit land area (see Chapter 3), dW/dt is average tree growth rate and we assume no weeds or understorey vegetation.

Equation (6.1) refers to a stand. If we were concerned with individual trees we would have to recognize that their rate of nutrient uptake varies with the population density. The growth rate of single trees may be vastly different in

6 Nutrient Dynamics and Tree Growth

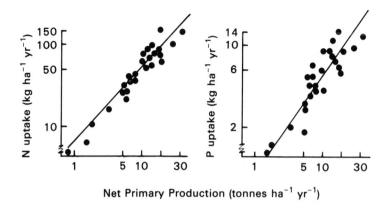

Net Primary Production (tonnes ha^{-1} yr^{-1})

Fig. 6.2. Uptake of nitrogen (N) and phosphorus (P) into above-ground components as a function of net primary biomass production, demonstrating that the amount of nutrient taken up is proportional to the dry matter increment (but note that the scales are logarithmic). The figures are redrawn from Miller (1984) who collated information from a number of sources. The data cover boreal, temperate and tropical forests.

full canopy stands of the same age but with different populations, or even within the same stand.

Nutrient uptake rate by trees, at any time, will be a function of soil moisture, temperature, the effectiveness with which the soil is exploited by roots (described by the parameter L_V; see Chapter 7) and—most difficult to quantify—the amount of "plant available" nutrient in the soil solution. The rate of dry matter production by the stand is dependent on intercepted energy and the photosynthetic characteristics of the foliage (see Chapter 5) and $[M]$ may vary depending on whether uptake rate is adequate to maintain it or not.

The average value $[M]$ is only useful in considering matters like nutrient turnover and the nutrient balance of a forest as a whole; in terms of individual trees we must consider the concentration in particular tissues and a more thorough description would be

$$[M] = \Sigma \, W_i [m]_i / W \qquad (6.2)$$

where $[m]_i$ is the mineral nutrient concentration in plant tissue i.

Equation (6.2) provides the basis for analysis of the "biochemical" cycle, i.e. nutrient translocation between plant parts. To illustrate: at time t the weight and nutrient concentration of the tissue i is characterized by $(W_i[m]_i)(t)$; if after an interval Δt, W_i has increased to $(W_i + \Delta W_i)$, but $[m]_i$ has remained constant then, since the product $W_i \times m_i$ (i.e. the total mass of nutrients) has increased, there must have been import of nutrients. As a numerical example, assume that biomass sampling at time t shows the foliage mass (W_f) of a stand to be 3 tonnes ha^{-1} with N-concentration of 1.5% (by weight); i.e. the total mass of N in the foliage of 0.05 tonnes, or 50 kg. A subsequent sampling, after interval Δt, shows that $W_f = 3.3$ tonnes ha^{-1}, but the N-concentration has not changed. Total mass of the nutrient has therefore increased to 55 kg. The calculation provides no information on whether the uptake came from the soil or from other tissues. This type of analysis will be considered in more detail later.

The term "plant-available" nutrients was used earlier. For mineral nutrients in the soil to be available to plants they must either be dissolved in the soil solution or held on exchange sites from which they can be absorbed by roots or mycorrhizal fungae. The concentration of these minerals is usually very low and although experimental work with nutrient solutions has shown that uptake rates increase with increasing solution concentrations, roots can remove nutrients from very dilute solutions (see Ingestad, 1982). The important factor determining the rate at which plants absorb nutrients is not so much the concentration of nutrients in the soil solution but the rate at which nutrients move to root surfaces. A thorough analytical treatment of water and nutrient movements to roots and root systems is provided by Nye and Tinker (1978). For present purposes it is enough to note that indications from research in this field are that, at least for agricultural crops, mobile ions such as NO_3–N move towards root systems by mass flow in liquid water as well as by diffusion along concentration gradients in the soil solution. Diffusion processes are generally slow, so that for immobile ions such as phosphate, uptake is limited by the volume of soil that can be explored by roots, and by root abundance. This should strictly be described in terms of absorbing surface area per unit volume of soil but, since this is almost impossible to measure, length per unit volume (L_V) is a good substitute. The extent to which soil is exploited by roots determines the length of the path along which nutrients must move to reach roots.

Since most nutrient uptake takes place from soil solution, it follows that uptake will be greatly reduced as soil dries. This may have important implications with regard to the effects of drought on tree growth. Nutrient concentrations and organic matter are usually highest in the upper soil layers, which dry most rapidly, so that reduction in nutrient uptake may be an early result of prolonged periods without rain. In fact reductions in nutrient uptake rates are likely to precede the development of serious water stress by a significant interval because tree roots may exploit the soil to quite considerable depth, where some nutrients are scarce but from which water can be extracted. This will result in maintenance of turgor and continued growth of aerial tissues, which would be expected to result in overall dilution of the mineral nutrient concentrations in trees.

The data of Fife and Nambiar (1984) support the suggestion that nutrient uptake is slowed by drought and also indicate that drought causes retranslocation from older to younger, more actively growing, tissues. There appear to be few other research data relating to nutrient retranslocation and uptake patterns by trees in relation to tissue water status.

The role of fungal mycorrhizae in the absorption of nutrients has been recognized since the turn of this century. The roots of virtually all tree species studied so far develop symbiotic associations with mycorrhizal fungi, which may affect growth and morphology and the exudation of organic substances from roots. Mycorrhizal stimulation of growth occurs most frequently in soils low in one or more nutrients (Bowen, 1985). Results presented in Table 6.1 illustrate the effects on the growth of Sitka spruce seedlings of inoculating two soils with several mycorrhizal isolates. All isolates increased growth although the effects were not consistent between the soils. Similar

Table 6.1. The effects on the growth of Sitka spruce (*Picea sitchensis*) seedlings of inoculating a nursery and a forest soil with different strains of mycorrhizal fungae. Data from Holden *et al.* (1983).

Soil	Fungus	Mycorrhizae as % of short roots	Dry weight (g)			Shoot/ root ratio
			Shoot	Root	Total	
Nursery	Control (uninoculated)	0	0.076	0.076	0.152	1.07
	E-strain	51	0.149	0.133	0.282	1.15
	Laccaria amethystina	81	0.380	0.421	0.801	0.95
	Unidentified (from nursery)	48	0.583	0.675	1.259	0.85
Forest	Control	0	0.020	0.016	0.036	1.43
	E-strain	69	0.164	0.142	0.306	1.49
	L. amethystina	84	0.218	0.194	0.412	1.20
	Unidentified	39	0.057	0.045	0.102	1.32

results from other authors are given by Bowen (1985). Increased uptake of nutrients occurs because ectomycorrhizal fungi act like extensions to the root system, in effect increasing root length density. This is particularly important in trees, where root length densities are low and, without mycorrhizae, exploitation of the soil volume is poor. The enhanced exploitation by mycorrhizae is particularly important for the uptake of immobile or poorly mobile nutrients.

The "classical" example of symbiotic relationships between plants and micro-organisms is N-fixation by Rhizobia. More recently an important relationship between Frankia and Casuarina species has been identified (see Bowen, 1985, and papers cited by him). In all the associations between beneficial mycorrhizae and higher plants it has been found that there are quite specific relationships between hosts and symbionts; particular strains of symbionts often appear to have evolved in association with particular provenances of trees. Knowledge of these relationships, and in fact of the whole complex ecology of mycorrhizal fungi and their association with trees, is still scattered and incomplete; a good compendium of recent papers is provided by the book edited by Atkinson et al. (1983). Anyone concerned with the establishment of trees, particularly in areas where the species they are concerned with has not been grown before, should try to obtain any information available about micro-organism associations with that species. The least that can be done is to inoculate seedlings with soil from successful growing areas.

Losses: fire. Fire is an important factor in the biogeochemical cycle and may be a major cause of nutrient loss from forest ecosystems (Walker et al., 1984), particularly in countries like Australia and the Mediterranean regions, where long dry periods may be followed by hot, low-humidity weather. The impact of fire in these regions is also likely to be greater, from the nutritional point of view, because the soils supporting forests are often poor.

Fires in forests are either wild (uncontrolled) or prescribed to reduce fuel loads and hence fire risk, or to assist site preparation and regeneration. The effects of hot fires are more serious than those of cool fires. It is obvious that fire releases nutrients from the organic matter more rapidly than would otherwise have been the case, but it must be emphasized that fires do not contribute any additional nutrients; they simply change their form, hasten some cycling processes and also result in losses of some nutrients. Nitrogen is volatilized during fire and its loss parallels the loss in weight of burnt fuel (Raison et al., 1985). Soil heating results in immediate release of significant amounts of NH_4-N as a result of oxidation of organic matter. Raison (1984) estimates that the quantity of N lost in burning the litter on a forest floor can equal or exceed the amount removed in harvesting. He gives figures of up to

6 Nutrient Dynamics and Tree Growth

1000 kg ha^{-1}. After a fire there may be nutrient losses by water erosion—ash is easily carried by surface run-off—or by wind removing nutrient-rich ash from the site. The nitrogen lost in fire must be replaced either by rain or by symbiotic or non-symbiotic fixation, and since addition rates by these processes are so slow (of order 0–10 kg ha^{-1} yr^{-1}), frequent burning of a forest will result in nitrogen depletion and probably reduction in the amounts of other nutrients available to the trees. Some phosphorus is also lost in volatile forms in fire; Raison (1984) estimates that 50% of the phosphorus contained in litter will be lost when the litter is burnt.

For those who wish to pursue the topic of the effects of fire on forest nutrition the reviews by O'Connel *et al.* (1981), Raison (1979), Rundell (1981), and Walker *et al.* (1986) provide valuable information and summaries of current knowledge.

Losses: tree harvesting. The other major source of nutrient loss from forests is tree harvesting. If harvesting is selective, with only a few trees removed, nutrient losses are small and can largely be made up by weathering and inputs from rain. However, when intensive plantation forestry culminates in clear-felling and the removal of large masses of timber the nutrient losses may be considerable. Various studies on the removal of nutrients by harvesting eucalyptus forests indicate losses of 10–30 kg P ha^{-1}, 30–160 kg K ha^{-1}, 100–600 kg Ca ha^{-1} and 20–100 kg Mg ha^{-1} (Attiwill, 1981). Data collated by Turner (1981) also fall within these ranges. He shows N-losses to be in the range 70–250 kg ha^{-1} for eucalyptus, and generally 200–250 kg ha^{-1} for *Pinus radiata*. Values for the other nutrients are similar in *P. radiata* and eucalyptus, and appear to be in the same ranges for other forest types (see Table 6.2). The nutrients removed by harvesting are in the stem and stem bark. Losses are exacerbated if the logging residue is burnt. Clear-felling and burning are likely to increase the rates of organic matter decomposition because of the disturbance—resulting in greater soil aeration and exposure, which may lead to higher soil temperatures. There are also likely to be increased losses by leaching of nutrients from the profile because there are fewer fine roots to absorb them in the period after clear-cutting—at least until herbs become established to utilize the nutrients and release them when they die and decompose.

Although there are now many data in the literature on the nutrients likely to be lost under intensive silvicultural systems, it remains important that there is continued research aimed at quantifying the rates of loss under different conditions. This is widely recognized, as witness the symposia devoted to this topic in recent years ("Impact of Intensive Harvesting on Forest Nutrient Cycling" at the School of Forestry, Syracuse, NY (1979); "Australian Forest Nutrition Workshop: Productivity in Perpetuity",

Table 6.2. Amounts of nutrients (kg ha^{-1}) in the stem wood, stem bark, living branch wood and living foliage of a 15-year-old *Pinus radiata* stand in Australia (data of Stewart *et al.*, 1981) and an aspen-mixed hardwood spodosol ecosystem in Wisconsin, USA (data of Pastor and Bockheim, 1984). These data indicate the potential losses that might result from clear-felling and burning; conversely the amounts of nutrients available for re-cyling under conservative management.

Component	Dry weight (tonnes ha^{-1})	N	P	K	Ca	Mg
			P. radiata			
Stem wood	109	89	14	100	56	24
Stem bark	14	37	4	42	34	13
Branch wood	16	65	8	68	56	16
Foliage	10.4	123	9	68	41	19
			Aspen–hardwood ecosystem			
Stem wood	124	97	13	190	160	25
Stem bark	24	86	11	83	400	17
Branch wood	23	92	14	78	220	17
Foliage	3.4	54	7	26	38	5

CSIRO, Canberra, 1981). These volumes contain many useful data on the nutrient composition of the biomass of various forests and forest types and the nutrient composition of the component parts (see Tables 6.2 and 6.3). Management for sustained yield from forests must include careful consideration, based on calculations of nutrient loss, addition and turnover rates, of the need to leave the largest possible amount of biomass on the ground and hence retain as much nutrient as possible. Such calculation may suggest the desirability of de-barking harvested trees, as well as the need to add fertilizers. In other words, management for sustained yield requires a good appreciation of the biogeochemical cycle, including the mineralization of the organic products of growth, the contribution of mycorrhizae to uptake of nutrients and to organic turnover, and the rate of supply of available nutrients from the soil.

6.1.3 The "Biochemical" Cycle

In the early stages of tree growth most of the nutrients taken up from the soil will be retained in the biomass, but as trees grow older the contribution of "biochemical" cycling to the nutrients required for growth increases. Before the formation of heartwood and while the foliage mass of trees is increasing, the rate of nutrient uptake and immobilization in tissue will be higher than the rate of release in litterfall or recycling in the biochemical cycle. At equilibrium, when the tree or canopy leaf area is stable, nutrient storage in new tissues is matched by the quantity of nutrients recycled, either through the "biochemical" or biogeochemical cycles. At this stage the "biochemical" cycle may be a major contributor of the nutrients needed for growth.

Biochemical cycling may be illustrated formally as follows. Assume that over some period Δt, the dry weight gain ΔW of a tree or stand can be calculated. ΔW is partitioned into leaves (ΔW_f), stems (ΔW_B) and roots (ΔW_r); we will assume, for present purposes, that heartwood has not yet formed, although it could easily be incorporated into this model. The nutrient concentrations in the tissues at time t, the beginning of the period of interest, are $[m]_f$, $[m]_B$ and $[m]_r$, where $[m]_i = m_i/W_i$ and average mineral nutrient concentration $[M]$ is given by equation (6.2). For example using data from the same sources as Table 6.2, and assuming the nutrient involved is phosphorus (so $m = P$): $(W_f[m]_f + W_B[m]_B + W_r[m]_r)/\Sigma W = (3.5 \times 0.003 + 150 \times 0.0001 + 20 \times 0.001)/173.5 = 0.000\ 26$, $[P]$ in this case being strongly affected by the weight of stems.

Dealing with the transfer processes sequentially (rather than simultaneously as happens in nature), tissue nutrient concentrations at time $(t + \Delta t)$ would be

$$[m]_i(t + \Delta t) = \frac{m_i(t)}{W_i(t) + \eta_i \Delta W} \quad (6.3)$$

where η_i is the partitioning coefficient to organ i (see Chapter 5). Over the interval Δt a mass of nutrient Δm_u is absorbed from the soil into the roots, a mass Δm_r is moved from the roots to stem and a mass Δm_B from stem to leaves. Therefore, at the end of the period, the tissue nutrient concentrations are

$$[m]_f(t + \Delta t) = \frac{m_f + \Delta m_B}{W_f + \eta_f \Delta W} \quad (6.4a)$$

$$[m]_B(t + \Delta t) = \frac{m_B + \Delta m_r - \Delta m_B}{W_B + \eta_B \Delta W} \quad (6.4b)$$

$$[m]_r(t + \Delta t) = \frac{m_r + \Delta m_u - \Delta m_r}{W_r + \eta_r \Delta W}. \quad (6.4c)$$

The consequences of various assumptions become immediately obvious; e.g. if $\Delta m_r < \Delta m_B$ nutrient concentration in the stem falls, and so on. The need for limits, knowledge of transport rates and feedback relationships also becomes clear: for example, what are the upper and lower limit values for $[m]_i$? Is η_i affected by $[m]_i$—particularly in relation to leaves (Linder and Rook (1984) suggest that it is) and roots? and so on. A set of sample calculations made with equations (6.7) is given in an Appendix to this chapter.

Miller (1984) presents data which show that between 50 and 60% of the requirements for new growth of mature *Pinus nigra* were met by retranslocation of major nutrients—mainly recovery from needles. Switzer and Nelson (1972) found that a "sizeable portion" of the annual nutrient requirements of 20-year-old loblolly pine (*Pinus taeda*) was met by transfers between the living fractions of the system, and that the soil was a surprisingly small contributor to total annual nutrient requirements (N, P, K, Ca, Mg, S). They found that only about 19% of nutrients were retained by the trees, the remaining 81% being recycled. Table 6.3 from Attiwill (1981) summarizes information on the concentration of various elements in green foliage and litterfall of various forest types and the withdrawal of elements from this foliage.

Attiwill (1980), studying mature *Eucalyptus obliqua*, found the concentration of phosphorus in the tissues to be low by comparison with other species. "Biochemical" cycling accounted for 46% of the gross annual P uptake by the stand. Attiwill found that a major contribution to "biochemical" cycling comes from heartwood formation; withdrawal of P prior to heartwood formation accounted for 31% of the total "biochemical" cycle of phosphorus and 17% of the "biochemical" cycle of potassium (Attiwill did not analyse for nitrogen). He considered that the formation of heartwood may be regarded as a growth-regulating process in which part of the annual net primary production is stored, rather than cycled as litter. The retranslocation of nutrients from heartwood in *Eucalyptus obliqua* is much higher than the recovery from tissues other than needles identified by Miller. It is clearly a matter of considerable importance.

Miller *et al.* (1976a) showed that nutrient concentrations in different foliage classes change markedly during the growing season and Fife and Nambiar (1982) demonstrated unequivocally that the nutrient concentration of foliage is subject to continual change. They found that the phosphorus content of *Pinus radiata* needles was 24.9 µg/needle after prolonged drought and 33.5 µg/needle three weeks after rain. Their results indicate that phosphorus was being withdrawn from needles (of any age) in favour of growing tissue when drought reduced or stopped uptake from the soil.

6.2 GROWTH IN RELATION TO NUTRITION

From the point of view of production forestry the most important questions

Table 6.3. Examples of element concentrations $[m]_f$ (g kg^{-1} dry weight) in green foliage (f) and foliage litterfall (l) in various forest types. Elements withdrawn from foliage (Δm_t) before litterfall were calculated as percentages of elements in green foliage (data collated by Attiwill, 1981).

Forest type		N $[m]_f$	Δm_t(%)	P $[m]_f$	Δm_t(%)	K $[m]_f$	Δm_t(%)	Ca $[m]_f$	Δm_t(%)	Mg $[m]_f$	Δm_t(%)
Hardwoods—Evergreen											
1. Australian eucalypts											
Wet sclerophyll forest	f	14.73		1.05		4.60		6.77		3.87	
	l	6.1	59	0.62	41	4.03	12	8.13	−20	2.77	28
Dry sclerophyll forest	f	11.8		7.32		3.42		4.54		3.10	
	l	5.95	50	2.72	63	2.31	32	6.18	−36	2.26	27
2. *Nothofagus truncata*	f	12.1		1.2		8.1		7.1		1.4	
	l	6.6	45	0.44	63	1.52	81	9.83	−38	1.76	−26
Hardwoods—Deciduous											
3. Mixed Quercus, Betula, and Fraxinus forest	f	24.41		1.32		11.08		11.94		2.71	
	l	15.45	37	0.61	54	5.29	52	18.22	−53	2.58	5
4. *Quercus robur, Tilia cordata*	f	16.2		1.04		6.0		6.0		2.0	
	l	14.74	9	0.60	42	1.37	77	8.98	−50	1.27	37
5. *Fagus sylvatica*	f	27.4		1.49		8.00		3.61		0.70	
	l	16.58	39	1.33	8	5.35	33	5.45	−51	0.49	30
Equatorial forest	f	25.92		1.62		12.72		18.99		1.63	
	l	14.64	44	1.22	25	10.74	16	7.34	61	3.51	−115
Softwoods											
6. *Pinus radiata*	f	13.2		1.8		9.0		1.2		1.13	
	l	7.9	40	0.77	57	3.1	66	5.2	−333		
7. *Pseudotsuga menziesii*	f	11.80		2.72		4.13		2.78			
	l	4.61	61	1.15	58	1.73	58	16.33	−487	0.24	79
8. *Picea abies*	f	10.39		1.20		5.10		9.70		1.40	
	l	9.19	12	1.09	9	4.40	14	9.07	6	1.45	4

in relation to nutrition are: is the rate of nutrient supply at a site adequate to ensure high and sustained yields; will the impact of forest management practices adversely affect subsequent nutrient supplies and growth; what returns, in the form of increased growth rate and greater final yield, might be expected from the application of a given quantity of fertilizer? In most regions the latter question will be virtually unanswerable unless there have been empirical fertilization experiments on the same soil type. Even if there have, they may not provide more than a general guide, particularly in areas of unreliable rainfall. Even in agriculture, after over 100 years of research on nutrition, it is not generally possible to make accurate predictions of the likely responses of crops to fertilization.

Our inability to predict quantitatively the effects of nutrient losses during management, or responses to fertilizer, results from both the complexity of the soil–plant system and the approach that has been adopted in much of the research in this area. An enormous amount of work has been done on various aspects of soil chemistry, aimed at characterizing soils in terms of their nutrient supplying ability. Much of this work has been associated with studies on plant response, but these studies have usually been of the correlative or "relational" type, i.e. total growth observed at the end of a period is related to soil properties, or to the quantity of nutrients supplied to the soil, or nutrient deficiencies are evaluated in relation to empirical plant response curves—again in terms of total growth. This type of empirical experiment does not lead to results of general application and, in the case of massive, long-lived plants like trees, it is difficult to do. There have also been many attempts to define critical foliage nutrient concentrations and to establish relationships between final yield, or growth increment over a period, and nutrient concentration in the foliage, measured at a particular time. These lead to highly variable results because tissue nutrient status at any time affects the growth rate at that time, and relationships between nutrient concentrations and amount of growth constitute an attempt to explain an integral (the state of the plant at the time) in terms of a variable affecting a rate (cf. Chapter 2).

A model. To describe growth as a function of nutrition we would expect to have to write models starting with the general relationship

$$dW/dt = f[M].$$

This would lead immediately to requirements for information (about functional relationships) that we do not have—information which would be very difficult to obtain. However, progress is possible using a different approach.

Ågren (1983) has proposed a model based on the concept of nitrogen

6 Nutrient Dynamics and Tree Growth 127

productivity—the amount of biomass produced per unit of nitrogen taken up by a stand. He dealt only with nitrogen because it is commonly one of the most growth-limiting macronutrients and relevant data are therefore readily available. Ågren also restricted his analysis to foliage, recognizing that growth of other tissues can be estimated from knowledge of foliage mass, distribution and photosynthetic properties (see Chapters 3, 4 and 5). The basic equation of this model is

$$\frac{dW_f}{dt} = \varepsilon_N N - \gamma_f W_f \qquad (6.5)$$

where N is the total mass of nitrogen in the foliage, ε_N is the nitrogen productivity (mass foliage/mass N yr^{-1}) and γ_f is the rate of foliage loss. Assuming γ_f to be constant and ε_N to decrease linearly with increasing W_f (it could not remain constant or foliage mass would increase indefinitely with added nitrogen) gives

$$\varepsilon_N = a_N - b_N W_f \qquad (6.6)$$

where a_N and b_N are species-specific constants.

For conifers with relatively stable, long-lived foliage biomass ε_N can be defined as (current foliage biomass)/(total nitrogen in all foliage). Ågren fitted equation (6.6) to a number of data sets, available in the literature, and obtained the results presented in Table 6.4.

Table 6.4. Values of the parameters a_N ((kg biomass)(kg N)$^{-1}$ yr^{-1}) and b_N (1/kg N yr^{-1}) of equation (6.6), from Ågren (1983). Note that Norway spruce (*Picea abies*) has lower intrinsic nitrogen productivity (low a_N) but the low b_N value indicates little reduction in N-productivity with increasing W_f. Nitrogen productivity is much higher for two of the pine species and intermediate for Douglas fir (*Pseudotsuga menziesii*) and Scots pine (*Pinus sylvestris*) but in all cases reduces relatively rapidly with increasing W_f. The accuracy of the estimates was not high.

Species	a_N	b_N
Picea abies	18	0.4×10^{-3}
Pseudotsuga menziesii	34	1.2×10^{-3}
Pinus nigra	50	1.8×10^{-3}
Pinus resinosa	56	2.4×10^{-3}
Pinus sylvestris	41	2.0×10^{-3}

If additional increments of nitrogen produce decreasing dry matter (foliage) increment, a point must be reached where $dW/dt = 0$, foliage mass is at its maximum ($W_{f.max}$) and the foliage contains the maximum possible amount of nitrogen (N_{max}). (Any further addition of N would either cause foliage loss, or increased foliage mass with reduced N-concentration.) Setting $dW/dt = 0$ in equation (6.5), substituting for ε_N from equation (6.6) and solving for N_{max} gives

$$N_{max} = \frac{\gamma_f W_{f.max}}{a_N - b_N W_{f.max}}. \tag{6.7}$$

The nitrogen concentration at this point is $(N_{max}/W_{f.max}) = [N]_{max}$ say, $= \gamma_f/(a - bW_{f.max})$. Whether this is the optimum N-concentration for growth is perhaps arguable, but we can use the relationship to arrive at Ågren's third equation:

$$W_{f.max} = \frac{a_N - \gamma_f/[N]_{max}}{b_N}. \tag{6.8}$$

Given estimates of a_N, b_N, $[N]_{max}$ and γ_f, equation (6.8) allows estimation of the maximum amount (mass) of foliage likely to be found in the forest type to which the parameter values apply. Ågren estimated γ_f, from the data sets used to obtain Table 6.4, as the ratio of current to total foliage biomass. He assumed that $[N]_{max}$ represented the optimal N-concentration for growth and obtained an estimate of it from the published results of laboratory experiments. Using the a_N and b_N values given in Table 6.4 he calculated values of $W_{f.max}$ that agreed quite well with observed values.

Ågren's (1983) analysis has been discussed in some detail because it represents significant progress towards the essential goal of quantitative description of the growth of trees in relation to nutrition. The model is written in terms of *amount* of the nutrient, rather than its concentration, but if we substitute the right-hand side of equation (6.6) for ε_N in (6.5) and divide through by W_f we obtain

$$\frac{1}{W_f} \frac{dW_f}{dt} = a_N \left(\frac{N}{W_f}\right) - (b_N N + \gamma_f) \tag{6.9}$$
$$= a_N [N]_f - (b_N N + \gamma_f).$$

This indicates that *relative* leaf growth rate depends on tissue nutrient concentration.

Ingestad (1982—and see the series of papers referred to in that paper) demonstrated, using nutrient solutions in the laboratory, that constant seedling relative growth rates, and constant internal nutrient concentrations, can only be maintained if the rate of nutrient supply to the roots (nutrient

6 Nutrient Dynamics and Tree Growth

flux density) increases in the same manner as growth; i.e. of the relative nutrient addition rate $((1/M)\,dM/dt)$ is constant

$$\frac{1}{W}\frac{dW}{dt} = c\frac{1}{M}\frac{dM}{dt}. \tag{6.10}$$

Equation (6.9) suggests that even if the uptake (addition) rate is fast enough to maintain [N] constant as W increases, $(1/W)\,dW/dt$ will decrease.

The equations presented above ((6.5)–(6.10)) are valuable for analytical purposes, and the insights they provide into growth in relation to nutrition. However, they are of limited value for predictive purposes unless the form of Ågren's model can be confirmed, and reliable parameter values established for many other species. (When appropriate data become available the analysis should also be tried for nutrients other than nitrogen.) Given the appropriate parameter values dW/dt could be estimated by the methods outlined in Chapter 5 and the nutrient requirement of the stand evaluated. To do this we will need to know how leaf growth and photosynthetic characteristics are affected by nutrition.

Linder and Rook (1984) recently reviewed the effects of mineral nutrition on carbon dioxide exchange and the partitioning of carbon in trees. They state categorically: "When a tree's growth is stimulated by fertilization a significant part, if not most, of the result is due to an increase in the total area of the photosynthetic surface", i.e. in leaf size and surface area and number of leaves produced. There are also likely to be effects on leaf longevity, although these may vary. In situations of serious nutrient deficiency, leaf numbers may be reduced by early shedding caused by nutrient withdrawal. This, of course, will cause increased litter production and may lead to immobilization on the forest floor. These conclusions indicate that the leaf partitioning coefficient (η_f, equation (5.1a)) is the factor most likely to be affected by nutrition and Linder and Rook conclude that fertilization of non-closed coniferous stands, where nutrient supply is limiting, leads to increased growth by increasing needle biomass. This leads to more effective energy interception as well as slightly higher photosynthesis rates at given photon flux densities. The effects of nitrogen on photosynthesis seem to be much more pronounced in hardwoods than in conifer seedlings but the limited information available on the overall effects of fertilization on hardwood stands did not allow Linder and Rook to draw conclusions about the effects on annual carbon balance.

6.3 CONCLUDING REMARKS

"Biochemical" cycling and the effects of periods of drought on the uptake of nutrients and their retranslocation could confuse the patterns of tree response to fertilizer application in relation to soil fertility. Miller (1984) says (somewhat teleologically) that internal ("biochemical") cycling enables "storage during periods of plentiful supply for use in possible future lean years. This explains much of the longevity of the response shown by trees to fertilizer application and their ability to show a fairly regular pattern of annual growth despite marked variations in soil nutrient supply." I believe that detailed analysis of nutrient dynamics is essential if we are to understand responses to fertilizer applications—including the differences between species and provenances—and arrive at rational and usable site classifications.

There is convincing evidence of genetic differences in the efficiency of nutrient utilization (dry matter produced/unit nutrient) by agricultural plants (Graham, 1984), and some evidence that there are differences between tree provenances (Cotterill and Nambiar, 1981). Differences in response of species and provenances to nutrients may be attributable either to differential response in terms of leaf photosynthetic properties, the carbohydrate partitioning factors (η_i) or to different patterns of nutrient translocation between component parts of the trees (the "biochemical" cycle). Species which allocate high proportions of their carbohydrates to root systems may be very effective in taking up nutrients, and hence respond well to improved nutrition. Investigating these processes, and, of course, continually attempting to identify the factors involved in the rate at which nutrients become available to roots (the "nutrient supply rate"), will provide more precise and useful information on the likely responses of forest stands to differences in soil nutrient status than the results of long-term field experiments, although such field experiments are, of course, necessary.

The "nutrient status" of sites is an important factor in site classification, the basic forestry tool used to estimate the potential productivity of sites, for particular species. This is usually done in terms of site indices, which are completely empirical and only provide quantitative estimates of tree growth when used in conjunction with locally derived (historical) yield tables. The methods available for measurement of soil nutrient status do not generally provide good estimates of the rate at which nutrients become available for use by plants, which is why it is so difficult to characterize sites on the basis of any form of soil analysis. Nevertheless, the capacity of sites to support growth must be defined in terms of some measure of the nutrient supplying power of the soil (e.g. soil depth exploitable by roots, long-term cation reserves and release rate (see Turner, 1981) combined with soil hydraulic characteristics, or dynamic indices such as rates of N-mineralization). Site classification should also include a term to account for the probable length of growing season—including accounting for the probability of drought periods and an average value for total incoming short-wave energy at the site (see

Tajchman, 1984). In effect then, site classification should take the form of a simple model of the potential growth at a site in terms of dry matter production and its average nutrient concentration [M]. To achieve this will require a much more mechanistic approach, along the lines suggested in this chapter, to research aimed at identifying the potential productivity of sites. There seems no reason not to use this approach—despite its greater complexity in relation to current methods—although a great deal of work in developing and testing models is needed. An outline of the type of model needed is given in Chapter 8.

APPENDIX TO CHAPTER 6

A set of sample calculations to illustrate "biochemical" cycling are given here, and briefly discussed.

We assume a stand with (at time t) the properties tabulated below:

Mass of foliage, $W_f = 5 \times 10^3$ kg ha^{-1}
Mass of stems, $W_B = 100 \times 10^3$ kg ha^{-1}
Mass of roots, $W_r = 20 \times 10^3$ kg ha^{-1}
$[P]_f$ (kg/kg) $= 0.003$
$[P]_B$ (kg/kg) $= 0.00012$
$[P]_r$ (kg/kg) $= 0.001$.

W_B is assumed to include stem and branch bark + wood, and no allowance has been made for differences between sapwood and heartwood. All values are "lumped" in this sense.

We assume a stand dry matter production rate of 100 kg ha^{-1} day^{-1} and $\Delta t = 10$ days, hence $\Delta W = 1000$ kg ha^{-1}. The partitioning coefficients (η_i, cf. Chapter 5) governing the allocation of assimilate to foliage, stems and roots respectively are

$$\eta_f = 0.2, \; \eta_B = 0.6, \; \eta_r = 0.2.$$

We assume that, after interval Δt, $[P]_f$ is unchanged, $[P]_B$ has reduced to 0.000 115 and $[P]_r$ is unchanged. The mass of phosphorus (P) in any tissue is $[P]_i W_i$.

Using equations (6.4) we can calculate the amounts of phosphorus (ha^{-1}) moved between compounds of the stand, and the amount that must have been taken up from the soil.

$$[P]_f(t+\Delta t)=\frac{P_f+\Delta P_B}{W_f+\eta\Delta W}.\quad (6.4a)$$

Under the assumption of constant $[P]_f$ after interval Δt, this gives

$$0.003=\frac{0.003\times 5\times 10^3+\Delta P_B}{5\times 10^3+0.2\times 1000}$$

therefore, $\Delta P_B=0.6$ kg, this being the amount of phosphorus that would have to be translocated from stems and branches to foliage to keep foliage P-concentrations constant while foliage mass increased by 200 kg.

$$[P]_B(t+\Delta t)=\frac{P_B+\Delta P_r-\Delta P_B}{W_B+\eta_B\Delta W}.\quad (6.4b)$$

Assuming that $[P]_B$ reduces from 0.000 12 to 0.000 115 kg P per kg wood and bark

$$0.000\,115=\frac{0.000\,12\times 100\times 10^3+\Delta P_r-0.6}{100\times 10^3+0.6\times 100}$$

therefore, $\Delta P_r=0.14$ kg, this being the amount of phosphorus that would have to be translocated from stems and branches to stems to maintain $[P]_B$ at 0.000 115 kg/kg. Note that if $[P]_B(t+\Delta t)$ was reduced far enough the calculations would indicate transfer from stems and branches to roots. It follows that there is a value of $[P]_B(t+\Delta t)$—in this case 0.000 113 kg/kg—at which there would be no need for nutrient transfer from roots to stems. In that case any uptake from the soil (Δm_u) would increase the nutrient concentration in the roots.

In the present example we assume that $[P]_B$ remains constant.

$$[P]_r(t+\Delta t)=\frac{P_r+\Delta P_u-\Delta P_r}{W_r+\eta_r\Delta W}$$

$$0.001=\frac{20\times 10^3\times 0.001+\Delta P_u-0.14}{20\times 10^3+0.2\times 1000}$$

therefore, $\Delta P_u=0.34$ kg, being the amount of phosphorus (ha^{-1}) taken up by the roots.

7 Water Relations

7.1 SCOPE

Plant–water relations is probably one of the most intensively studied areas in plant physiology and plant–environment interactions. Slatyer's (1967) book was a landmark in the field but there have been many books and major reviews since, some devoted exclusively to the water relations of trees. Among these the monograph "Water Deficits and Plant Growth" (1981), Volume VI in the series edited by Kozlowski, is concerned with woody plant communities and contains several valuable chapters, while the booklet by Hinckley et al. (1978) provided a comprehensive survey of current knowledge on temporal and spatial variations in the water status of forest trees. Kozlowski (1982) has recently published a review which is essentially a catalogue of all recent papers concerning the water relations of trees; he does not attempt to synthesize principles. Other important papers are cited in this chapter.

Although there is obviously much duplication in all this literature, it does reflect the mass of information at all levels in the field. In this chapter I shall outline the theoretical framework within which analyses of plant–water relations are usually set—beginning with a treatment of the parameters used—and then discuss the consequences of water stress for tree growth and productivity. The objective is to provide the principles necessary for understanding, at the level of processes and mechanisms, the results obtained in experiments or observations made in the field, so that these can be translated into reliable predictions of the consequences of change at tree and stand level.

Tree–water relations provide an excellent and well-documented example of processes at different levels with different response times. I have already discussed (in Chapter 3) leaf energy balance and its interaction with stomata. Response times to changes in leaf energy balance are of the order of seconds, leading to rapid changes in the transpiration rate of individual leaves. When water is lost by transpiration from the mesophyll cells adjacent to the sub-stomatal cavities of a leaf it is replaced by water moving from surrounding tissue, in turn replaced by water moving into the leaf tissue from the plant's conducting system. Changes in the water content of tissue cause changes in tissue water potential (see 7.2), resulting in potential gradients through the plant and from the soil to the roots. These gradients provide the driving force for water uptake from the soil and movement through plants. The relations between rates of transpiration and rates of water uptake and movement through the soil–plant system, across the frictional resistances to flow through the system, therefore determine the water status of plants.

Although it is relatively easy to measure changes in plant–water status and processes such as photosynthesis over periods of minutes, processes such as cell division and expansion, which are strongly affected by tissue water status, can only be satisfactorily measured in intact plants over much longer periods. The end results of these processes appear as shoot and leaf growth and stem extension. These can be easily measured over periods of hours, although significant trends are not likely to be clearly detected over periods less than days. Longer-term changes associated with water stress include leaf development and growth in leaf area, production of cambium and xylem in stems, and root growth. Some of these will be discussed in more detail later.

Calculations of the soil water balance, using rainfall data, stand evaporation and transpiration rates and estimates of run-off and drainage (see equations (7.29) and (7.30)) can easily be made over hourly intervals. Calculations of transpiration rates can be checked by physiological and micrometeorological measurements but, because of the variability in forest ecosystems, soil–water balance calculations cannot be checked by direct measurement over periods of less than one to two weeks. This is probably the realistic interval to use for evaluation of the effects of water status on growth patterns and productivity at the stand level.

This chapter provides information on the parameters used to define water status, on what constitutes the rooting volume and on water movement through trees. The last sections deal with the hydrological balance at stand level and the consequences of water stress for growth, including concepts such as water use efficiency. There is some discussion of adaptation to water stress and its consequences.

7.2 WATER RELATIONS PARAMETERS

The most widely used parameter in water relations studies is tissue water potential (ψ, MPa—i.e. it has units of pressure, or energy per unit volume, equivalent to a force per unit area). Water potential is soundly based in thermodynamic theory (see Slatyer, 1967; Passioura, 1982, for a discussion of an area where confusion has arisen; and Tyree and Jarvis, 1982). Since water potential gradients indicate different energy states, the concept is fundamental to analysis of water movement through the soil plant system. However, water potential is not necessarily the best parameter to use in analysis of physiological functions. Others that may be more pertinent are relative water content (R^*) and turgor pressure—a component of ψ. Relative water content is the ratio of the weight of water held in tissue to that which can be held by the tissue at full turgor (see equation (7.4)). The inter-relationships between total water potential, turgor potential (or hydrostatic pressure, ψ_p), osmotic potential, or osmotic pressure (ψ_π), are given by equation (7.1):

$$\psi = \psi_p + \psi_\pi. \tag{7.1}$$

Equation (7.1) omits the so-called matric potential, thus implying acceptance of Passioura's (1982) argument that matric potential—which is supposed to account for the influence of solids on ψ, arises because of an inconsistent definition of pressure.

The turgor potential term provides a measure of the turgor pressure in cells (in tissues). ψ may vary across tissue but at equilibrium it will be approximately uniform; ψ_π and ψ_p may vary but equation (7.1) will hold. The turgor potential is usually estimated by the difference between measured total potential and osmotic potential. Total water potential in leaves and twigs is generally measured with a pressure chamber. Osmotic potential depends on the concentration and species of solutes in the cell sap and is given by the van't Hoff equation:

$$\psi_\pi = -\Phi RT\rho_w C$$
$$= \Phi RT(N_s/V) \tag{7.2}$$

where R is the universal gas constant (8.31 J mol^{-1} K^{-1}), T is Kelvin temperature, ρ_w is the density of water, $C = N_s/w$—the total number of moles of solute per unit mass of water in the symplasm (cell water volume V), and Φ is an osmotic coefficient that accounts for the fact that the solutes are non-ideal in a thermodynamic sense. Tyree and Jarvis (1982) provide a detailed discussion of ideal and non-ideal solutions, and osmotic potential, which can be measured by freezing point or psychrometric techniques. They recommend measuring ψ_π at temperature T by a psychrometric technique.

The water content of cells (and tissues) can be related to ψ by the so-called Höfler diagram (see Fig. 7.1; also Hellkvist et al., 1974; Tyree and Jarvis, 1982) which relates the volume of water remaining in the tissue to the total and constituent potentials acting on it (see equation (7.5)). This can also be derived from a pressure–volume curve. Essentially the method consists of measuring the water potential of tissue that has been brought to a series of different relative water contents. This can be done in a number of ways; one is to progressively raise the pressure in a pressure chamber and measure the amount of water expressed from the cut end of the twig or leaf when the system has come to the balance point. As water is lost from the tissue the average value of ψ_p in the tissue tends towards zero, and at lower potentials $\psi = \psi_\pi$ (see equation (7.1)). From equation (7.2), at any potential

$$\frac{1}{\psi_\pi} = \frac{-w_s}{\Phi RTN\rho_w} \qquad (7.3)$$

where w_s is the mass of symplasmic water remaining in the tissue. Plotting the reciprocal of pressure applied against the mass of water expressed yields a curve like that shown in Fig. 7.1a; the point where the curve becomes linear is

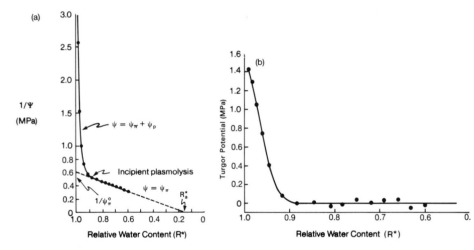

Fig. 7.1. (a) A "pressure–volume" curve, produced by plotting the reciprocal of pressure applied to a Eucalyptus globoidea leaf (1/ψ) against relative water content (R*). Extrapolation of the linear region to the 1/ψ axis provides an estimate of the osmotic potential of the tissue at full turgor, in this case −1.66 MPa. Extrapolation to the R* axis gives an estimate of R* for the apoplasmic (bound water) fraction of the tissue (15.7%). (b) The relationship between tissue turgor potential (ψ_p) and R* for the same leaf. This is derived by calculating the tissue osmotic potential (ψ_π) corresponding to each value of R*, using the linear equation describing the straight-line portion of the $\psi^{-1}/R*$ relationship. Each (calculated) value of ψ_π is subtracted from measured ψ to obtain ψ_p (see equation (7.1)). This figure illustrates clearly that, for this leaf, $\psi_p = 0$ when R* = 88.4%. (These data were kindly provided by Brian J. Myers, CSIRO Division of Forest Research.)

7 Water Relations

the point of incipient plasmolysis and extrapolating the linear portion backwards to the $1/\psi$ axis provides an estimate of the osmotic potential of the tissue at full turgor.

In practice, pressure–volume curves are usually expressed in terms of relative water content (R^*), which allows for different leaf sizes and provides a valuable measure of tissue water status. R^* is given by the expression

$$R^* = \frac{\text{weight of water in tissue}}{\text{weight of water in tissue at full turgor}}$$

i.e.

$$R^* = \frac{w_s + w_a}{w_s^0 + w_a^0} \quad (7.4)$$

where w_a is the apoplasmic water and w_s^0 and w_a^0 are the masses of symplasmic and apoplasmic water at full turgor. Substituting for w_s from equation (7.4) in equation (7.3) gives

$$\frac{1}{\psi} = \frac{w_a - R^*(w_s^0 + w_a^0)}{\Phi RTN\rho_w}. \quad (7.5)$$

Provided w_a can be taken as constant, the osmotic potential at full hydration is found by extrapolation of the line to $R^* = 1$ where $1/\psi = -w_s^0/\Phi RTN\rho_w = 1/\psi_\pi$. w_a^0 is then found from the intercept on the R^* axis and the relationship between ψ_p and R^* is derived from the pressure–volume curve, from the difference (see Fig. 7.1b) between the curves for $1/\psi$ and $1/\psi_\pi$. For the *Eucalyptus globoidea* leaves that provided the data in Fig. 7.1, osmotic potential at full hydration was -1.66 MPa, turgor potential was zero at $R^* = 88.4\%$ and w_a^0 constitutes 15.7% of the water in the leaves.

The technique of progressive measurement of expressed water is slow and tedious (although there are currently a number of variants and improvements) but pressure–volume curves provide an invaluable method of comparing the water status of different species, or the same species at different times of the season or grown under different conditions. Once the relationships between ψ and R^* have been established for particular tissue a single measurement of ψ on similar tissue will provide estimates of ψ_p, ψ_π and R^*.

7.3 THE ROOTING VOLUME

The effectiveness of water uptake by roots depends on the effectiveness with which the soil is exploited by roots (defined by root length per unit volume,

L_V, see Chapter 6), contact between roots and soil and the potential gradient between roots and soil. This gradient depends on the rate of water loss from the plant and the soil moisture content and hydraulic characteristics.

The factors affecting the mass of roots present under forest communities have been discussed briefly in Chapter 5. Equations given by Jackson and Chittenden (1981), from which the root mass of *Pinus radiata* can be estimated from linear dimensions, were presented in §5.4. Gerwitz and Page (1974) collated data on root length distribution under crops and found that in virtually every case L_V declined exponentially with depth (z), i.e.

$$L_V(z) = L_V(0) \exp(-k_L z). \qquad (7.6)$$

The rate of decline is defined by the constant k_L. Some data for forests can be analysed in this way. Carbon et al. (1980) examined the distribution of roots under jarrah (*Eucalyptus marginata*) in Western Australia, in soils with a sandy upper layer 1 m deep, overlying about 2–2.5 m sandy loam and up to 20 m of clay. They present two representative profiles derived from sampling at 25 sites. These show exponential reduction in $L_V(z)$ from $L_V(0) \approx 10^4$ m m^{-3} through the sand and the sandy loam, with an indication of a different rate of change (k_L) through the sandy loam in one of the profiles. In both cases large vertical roots penetrated to the 20 m limit of their sampling. Carbon et al. comment on the "almost total dominance of root length by fine to very fine roots".

In general it is probably safe to assume that changes in root length per unit volume with depth in forests can be described by equation (7.6); variation under forests and the errors in determination are such that there is little purpose in speculating about more sophisticated descriptions, which could not be tested. Doley (1981) tabulates data for rainforest and evergreen forests, on different soil types, which illustrate the point. Information on total root mass can be converted to estimates of root length per unit volume by appropriate constants (which will undoubtedly vary with species) for each class of roots (coarse, fine, very fine; see Chapter 5). Estimates of k_L can be made by specifying the level at which $L_V(z) \approx 0.05 L_V(0)$, hence $k_L z = 3$ and k_L can be estimated (e.g. if $L_V(1.0)/L_V(0) = 0.05$, $k_L = 3$).

The interactions between L_V and transpiration rates—and hence water uptake rates and soil hydraulic characteristics—are discussed in more detail in the section on "Water movement through trees". As a basis for this discussion a brief summary of soil hydraulic properties and the processes of water flow through soils follows.

The soil water potential (ψ_s) is the sum of gravitational, pressure or matric, and osmotic potentials. In the case of soils the matric potential is a very important component, being a function of soil particle size and arrangement.

7 Water Relations

Soil wetness is usually expressed in mass or volume terms; water content (mass water per unit mass soil) is related to volumetric water content (volume water per unit volume of soil θ_s) by the soil bulk density (ρ_s) and the density of water (ρ_w).

Soil water content and potential are related by the soil moisture characteristic curve (see Fig. 7.2), which can be described over much of the range by the empirical relation (Gardner et al., 1970).

$$\psi_s = a_s \theta_s^{-n_s}. \qquad (7.7)$$

Landsberg and McMurtrie (1984) give a complex expression for calculating

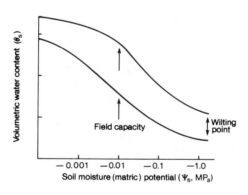

Fig. 7.2. Soil moisture characteristic curves for a heavy (upper curve) and a light (lower curve) soil. The ψ_s-scale is logarithmic. Soil moisture potential is called matric (suction) potential because the determinations were made using pressure membrane apparatus and do not reflect any osmotic components. Note the massive differences in water content at the same suctions (redrawn from Greacen and Hignett, 1984).

the effective soil water potential in the root zone ($\bar{\psi}_s$) from the soil water content/water potential relationship and root length density. Jones (1983a) has described an experimental method of estimating $\bar{\psi}_s$ in dry soil from measurements of leaf water potential (ψ_l) in trees in dry and wet soil. The two methods have not yet been compared.

Water movement through soil depends on the difference in ψ_s between any two points in the soil—and on the hydraulic conductivity, K_s. It can be described by Darcy's law:

$$J_s = K_s \frac{d\psi_s}{dx} \qquad (7.8)$$

where J_s is the volume flux of water through unit cross-sectional area per unit time (flux density) and x is distance. Darcy's law was originally formulated to describe flow through saturated media. Hydraulic conductivity falls rapidly as soil wetness falls below saturation, hence if equation (7.8) is to be used to describe flow in unsaturated soil K_s must be made a function of the water potential, i.e. $K_s = K_s(\psi)$.

Over the range of interest in plant studies the soil water potential varies from about 0 to -2 MPa. K_s may be reduced by several orders of magnitude across this range. The expression

$$K_s(\psi_s) = c\psi_s^{-n} \qquad (7.9)$$

(Gardner, 1958) provides a generally useful approximation to the relationship. Values for n varying from 2 to 4 have been found; they should be established experimentally for particular soils. Values of K_s for sand (0.28 m day$^{-1} \approx 3.3 \times 10^{-6}$ m s^{-1}), sandy loam (0.022 m day$^{-1} \approx 2.5 \times 10^{-7}$ m s^{-1}) and clay (0.01 m day$^{-1} \approx 1.2 \times 10^{-8}$ m s^{-1}) at a range of values of ψ are given by Carbon et al. (1980). As soil dries $K_s(\psi)$ can be expected to fall as low as 1×10^{-12} m s^{-1}.

Water movement through soil may also be treated in terms of the hydraulic diffusivity, D_s which is the ratio of the hydraulic conductivity K_s to the specific water capacity ($\partial\theta_s/\partial\psi_s$) (see Hillel, 1980a):

$$J_s = -D_s(\theta_s)\, \partial\theta_s/\partial x \qquad (7.10)$$

with

$$D(\theta_s) = a_w \exp(b_w \theta_s). \qquad (7.11)$$

$D_s(\theta_s)$ is a more conservative parameter than $K_s(\psi)$, varying from about 1×10^{-3} to 1 m^2 day^{-1} (i.e. 1.2×10^{-8} to 1.2×10^{-5} m^2 s^{-1}).

Much more detailed treatment of these relationships can be found in any good soil physics text, such as that of Hillel (1980a). The numerical values of the parameters in many of these physical soil–water relationships are conventionally determined in homogeneous, sieved samples of soil. They may provide very inaccurate indications of the real situation in the highly variable, heterogeneous environment in which roots function in the field. However, the equations do provide a guide to the type of relationships that can be expected.

7.4 WATER MOVEMENT THROUGH TREES

We have already noted that water moves from soil to roots and through plants along potential gradients, caused primarily by changes in leaf water potential as a result of transpiration. The literature on water movement through plants has, in recent years, been dominated by treatments in terms of the Ohm's law analogue which follows from this gradient-driven flow. Using this analogue the flow of water (J, m^{-3} s^{-1}) through the tree–soil system can be represented by flow through a series of hydraulic resistances, and described by the equation:

$$J = \frac{\psi_s - \psi_f}{R_s + R_r + R_x + R_f} \quad (7.12)$$

where R_s denotes resistance to water flow through the soil to the root, R_r resistance through the root to the xylem, R_x resistance to flow through the xylem and R_f from the xylem to the evaporating surfaces in the leaves.

If equation (7.12) holds it implies that there must be flow continuity and mass conservation through the system, so that a given volume of water ($J\Delta t$) lost from the leaves in time interval Δt will result in the extraction of the same volume of water from the soil over that interval. The relationship between J and $\Delta\psi$ becomes non-linear if the resistances are not constant, and this has been found to be the case in some experiments on herbaceous plants (see Passioura (1982) for a discussion). However, for trees in the field the possibility of non-constant resistances to flow through the plants can probably be safely ignored until it is shown unequivocally to be important. In a particular soil R_s will vary with θ_s (see equations (7.8) to (7.11)) and with L_V (see later discussion).

If the resistances are constant and equation (7.12) holds it can be easily solved by rewriting it as

$$\psi_f = \psi_s - J \Sigma R_i. \quad (7.13)$$

This is essentially an equation for a tree with a single leaf (at potential ψ_f) to which all the water flows. More rigorously, it would be necessary to subtract the summed products of the partial flows and resistances, which change in a branched system. This is discussed by Richter (1973). However, equation (7.13) may provide a good first approximation to the average value of ψ_f. If it does, plotting ($\psi_f - \psi_s$) against J should yield a straight line with a slope (ΣR_i), giving the sum of the resistances in the flow pathway. If the soil is wet ψ_s can be taken as approximately zero, so the plot reduces to ψ_f against J (or against transpiration rate which may be taken as an estimate of J). Equation

(7.13) has on occasion been adequate for this type of analysis (Landsberg *et al.*, 1975) but in most cases, particularly if the diurnal course of transpiration is plotted against (say) hourly average values of ψ_f such plots yield a hysteresis loop (Fig. 7.3). This appears to be a consequence of the fact that trees may store significant quantities of water in tissue such as sapwood (Landsberg *et al.*, 1976; Waring and Running, 1978).

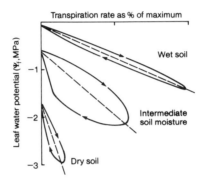

Fig. 7.3. Diurnal course of leaf water potentials as a function of (normalized) transpiration rates for trees in varying soil moisture conditions (redrawn from Hinckley *et al.*, 1978). The arrows in the curves denote their direction in time (falling in the morning, rising in the afternoon). As soil water potentials fall, falling potentials in the flow pathway result in increasing flows out of (tissue) storage. As transpiration rates decline there is replacement (see equations (7.26) and (7.27)). Under the driest soil moisture regime transpiration is greatly restricted. The broken lines represent the slopes (resistances) that would result from fitting equation (7.13) (see also Landsberg *et al.*, 1976), so the intercept on the ψ_f axis provides an estimate of the "effective" soil water potential.

Figure 7.3 shows relationships between transpiration and leaf water potential at different values of ψ_s. It can be interpreted as follows: as the transpiration rate increases during the first part of the day, water is withdrawn from the soil and from storage in tissue. In non-saturated soils water will have moved, during the night, to re-wet the regions around roots, dried out during the previous day. The extent of the drying out, the distance the water has to move, the wetness of the soil and its hydraulic properties (equations (7.9)–(7.13)) all influence the extent to which the soil surrounding the roots is recharged. We would expect that before transpiration commences ($J = 0$) leaf, xylem and soil water potential would all be approximately equal (pre-dawn water potential). If xylem water potential is higher than ψ_s there could be water movement from plant to soil, although Landsberg and Fowkes (1978) showed that there is no need to invoke some hypothetical one-way valve to stop this; it will always be negligible. Pre-dawn water potential is a useful and widely used measure of the equilibrium point between plant and soil water potential, although the night period is frequently not long enough for equilibrium to occur.

As water is withdrawn from the roots and from storage tissue at high transpiration rates, potentials in both these regions fall so that to sustain a particular flow rate later in the day leaf water potentials must fall further than earlier in the day. When transpiration rates fall—and hence leaf water potentials rise—later in the day, water moves from the transpiration stream back into storage tissues.

Models including storage. It was, at one time, argued that these phase lags between transpiration rate and ψ_f were a consequence only of the differences in rate of water movement to the roots through the soil and through the stem. However, numerous studies on lags between ψ_f and stem shrinkage, and patterns of swelling and shrinkage in cambium (well reviewed by Whitehead and Jarvis, 1981) confirm that equation (7.13) is inadequate to describe the patterns of water movement through trees (and probably through all plants; see Jones (1978); Wallace *et al.* (1983); Wallace and Biscoe (1983) for studies on wheat). It therefore becomes necessary to include the fluxes in and out of storage in models that predict the time course of leaf and other tissue water potentials. This makes it necessary to specify the relationships describing the water content/water potential relations in tissues.

Whitehead and Jarvis (1981) provide a detailed exposition of the relationships between the various parameters describing water in tissues. The most immediately useful of these is

$$v_o = V_s(F_{vs} - F_{bs}) \tag{7.14}$$

where v_o is the volume of freely available water in the sapwood volume V_s, F_{vs} is the volume fraction of water ($=v_o/v_s$) and F_{bs} is the fraction of bound water—a proportion (B_s) of the volume fraction of water in saturated wood (F_{ss})—which must be estimated experimentally. v_o can also be expressed in terms of the densities of wet wood (ρ_{fs}), oven dry wood (ρ_{ds}) and water;

$$v_o = v_s\left(\frac{\rho_{fs} - \rho_{ds}}{\rho_w} - B_s F_{ss}\right). \tag{7.15}$$

Values of ρ_{ds} and ρ_{fs} range between about 300–600 and 800–1100 kg m^{-3}, respectively.

Waring and Running (1978) use the expression

$$v_o = v_s \left(1 - \frac{\rho_{ds}}{1530}\right) \quad (7.16)$$

where the numerical value (1530 kg m^{-3}) is the density of solid material in wood, which can be regarded as constant (Siau, 1971; Skaar, 1972). Estimates of the total amount of "available" water stored in the foliage, cambium and phloem, sapwood and roots of forest stands are collated by Whitehead and Jarvis (1981). The commonest value was equivalent to about 0.4 mm depth, sufficient to supply transpiration for periods of up to several hours.

The change in water content of tissue per unit change in water potential (i.e. the slope of the pressure/volume curve) can be defined as the capacitance (Landsberg et al., 1976);

$$C = V_t \, d\theta_t / d\psi \quad (7.17)$$

where θ_t is the volumetric water content of the tissue, i.e. volume of water per unit volume of tissue (V_t). It follows from (7.17) that the rate of change of volumetric water content of tissue for unit change in water potential is

$$V_t \frac{d\theta_t}{dt} = C \frac{d\psi_t}{dt} = -J_t \quad (7.18)$$

where $-J_t$ is the flux out of the tissue in question.

Equation (7.12) can be broken into separate linear equations describing flux from soil to roots, through the roots to the xylem, through the xylem and from the xylem to the evaporating surfaces in the leaves. The equations that follow describe the flux from the soil into the xylem conducting tissue of the roots and stem (equation (7.19), which incorporates resistance to flow through the roots), and the fluxes from stem and leaf storage into the transpiration stream (equations (7.20) and (7.21)). Fluxes from storage, causing changes in water content, would be expected to cause changes in tissue volume. Assuming these to be small (equations (7.23) and (7.24)) allows simplifications that lead to expressions for the time course of leaf water potential and storage tissue water content (equations (7.26) and (7.27)). The equations developed here would apply to a single tree, or to a stand for which the parameters would be "bulked"—i.e. they would refer to unit land area, not to individual plants.

Flux from the soil into the xylem conducting tissue of the roots and stem is given by

7 Water Relations

$$J_x R_s = (\psi_s - \psi_x) \tag{7.19}$$

while flux into the xylem from trunk (plus branch, plus root) storage is given by

$$J_c R_c = (\psi_c - \psi_x) \tag{7.20}$$

where ψ_c is the water potential in storage tissue and R_c is the resistance to flow between storage and xylem tissue. Total flux from the xylem to the leaves is therefore

$$(J_c + J_x) R_f = (\psi_x - \psi_f) \tag{7.21}$$

and if total flux from the leaves (transpiration rate per unit leaf area $(E_t) \times$ leaf area (A_f)) is J_f then the change in leaf water content is

$$\frac{d(V_f \theta_f)}{dt} = (J_c + J_x) - J_f \tag{7.22}$$

The bracketed term on the right-hand side is the influx to the leaves. If changes in the volume of storage tissue (V_c, probably mainly sapwood) can be regarded as small then using (7.20) we obtain

$$V_c \frac{d\theta_c}{dt} = -J_c = \frac{\psi_x - \psi_c(\theta_c)}{R_c} \tag{7.23}$$

and assuming that $J_c \ll J_x$, and that changes in leaf volume are negligible, (7.22) gives

$$V_f \frac{d\theta_f}{dt} = J_x - J_f \tag{7.24}$$

The same assumption (that $J_c \ll J_x$) allows elimination of J_c from (7.21) and, combining (7.21) and (7.19) we obtain

$$(\psi_s - \psi_f) = J_x (R_s + R_f) \tag{7.25}$$

which is a steady-state approximation consistent with equation (7.12). Solving for J_x in (7.25) and substituting into (7.24) gives

$$V_f \frac{d\theta_f}{dt} = \frac{\psi_s - \psi_f(\theta_f)}{R_s + R_f} - J_f \tag{7.26}$$

and using (7.19) and (7.25) to eliminate ψ_x from (7.23) results in

$$V_c \frac{d\theta_c}{dt} = \frac{1}{R_c}\left(\frac{R_s\psi_f + R_c\psi_s}{R_s + R_f}\right) - \frac{\psi_c(\theta_c)}{R_c}. \quad (7.27)$$

Equations (7.26) and (7.27), written in finite difference form, provide the basis for numerical simulation of the time course of leaf water potential and storage tissue water content. Estimates of all the resistances are available and can be obtained by the methods used by Landsberg *et al.* (1976), Powell and Thorpe (1977) and Running (1980). The relationships between ψ_c and θ_c have been determined for *Pseudotsuga menziesii* by Waring and Running (1978) and for *Pinus sylvestris* by Waring *et al.* (1979); they are of the form $\theta_c = \text{constant} \times \theta_c^{-n}$ (see equation (A.7.3)). Similar relationships must be determined experimentally for other species. The simulation procedure is illustrated by a detailed worked example in the Appendix to this chapter, which serves to illustrate the information and relationships required and how they can be used. It also provides indications of the numerical values of the parameters.

An alternative model of water movement through trees is illustrated in Fig. 7.4. It depends on the assumption that water movement through a stem can be described by Darcey's Law written in the form

$$J = K_c(\theta) \frac{d\psi_c(\theta)}{dz} \cdot \frac{A_B}{\eta'}. \quad (7.28)$$

η' is the viscosity of water (N s m^{-2}) and A_B is stem cross-sectional area. The model may be solved by using finite differences and considering the tree (height $z = h$ at the base of the canopy) as being divided into any number of segments ($i = 1, n$) of length Δz. The information required is $K_c(\theta)$—see Waring and Running (1978)—and $\psi_c(\theta_c)$. Calculations begin with starting values of $\theta(z)$ and hence ψ_c; and should include the gravitational component $\rho_w Gz$, where G is acceleration due to gravity. The first step removes $J\Delta t$ m^3 water from the top segment, hence $\theta - J\Delta t = \theta(t + \Delta t)$ and ψ_c goes to $\psi_c(t + \Delta t)$, creating a gradient $\Delta\psi/\Delta z$ between the mid-point of the top segment ($i = 1$) and the next one down ($i = 2$). The hydraulic conductivity $K_c(\theta)$ is taken to be the value calculated from $(\theta(1) + \theta(2))/2$, and water moves from segment 2 to 1 along the gradient, for the period Δt. This creates a gradient from segment 2 to segment 3—all parameters are updated and the calculations continue. Movement from the soil into the base of the stem is across the difference $\psi_s - \psi_n$ through the resistance R_s. Simulations using this model give the results shown in Figs 7.5 and 7.6, which clearly illustrate the lags and differences in flow through the upper and lower stem, caused by tissue capacitance.

The soil–root resistance. As noted earlier root–soil resistance R_s—the

7 Water Relations

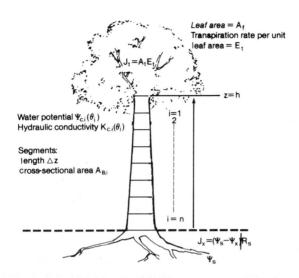

Fig. 7.4. A model of water movement through trees based on the assumption that the water lost in interval Δt is removed from stem segment 1, causing a change in water potential ($\psi_{c,i}$) and hence movement of water from segment 2 to segment 1, according to equation (7.28). Hydraulic conductivity ($K_c(\theta_c)$) is a function of stem water content. The water content of all segments is updated in each interval. Flow into the bottom segment is across the root–soil resistance R_s (see equation (7.19)). The non-linearity of the relationship between K_c and θ_c causes lags in the changes in water potential with height. The model can also be written to include storage in leaves and branches, as well as stem segments, provided their storage capacity can be specified (see Figs 7.5 and 7.6).

resistance to flow from soil to roots—depends on root length, soil wetness and hydraulic properties and the contact between roots and soil. There have been many analyses of the process of water movement to roots and mathematical evaluations of the consequences of varying L_V (see Hillel, 1980b). At one time it was considered that the main source of resistance was the reduction in soil hydraulic conductivity caused by extraction and the consequent dry zones around the roots, but the general consensus that has emerged from theoretical and experimental studies is that resistance to flow through roots themselves may be the dominant term when soil is wet, while low soil hydraulic conductivity becomes increasingly important as soil dries. A further factor which may be important is root–soil contact—the length of root in close contact with the soil may decrease as soil dries (for an experiment with trees see Dosskey and Ballard, 1980). Root–soil contact may may also be poor simply because of the heterogeneous nature of soils and the fact that roots tend to grow preferentially down paths of least resistance—

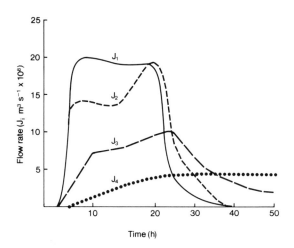

Fig. 7.5. Simulated flow patterns from branches to leaves (J_1), stem to branches (J_2), through the stem (half-way up, J_3) and into the stem at ground level ($J_4 \simeq J_x$). Flow out of the leaves (J_l) was taken as constant ($\simeq 20 \times 10^{-8}$ m^3 s^{-1}) from hours 4 to 24 (redrawn from Whitehead and Jarvis, 1981).

such as cracks and holes—where root–soil contact may be intermittent. Landsberg and Fowkes (1978), in their analysis of water movement through plant roots, and the source of resistances to flow, identified "effective root length" as a factor in root resistance and Herkelrath et al. (1977) incorporated it into a comprehensive uptake model. Very simplistically the consequences of a change in effective root length may be illustrated using equation (7.19).

Dividing the flux (J) by the effective root absorbing surface (A_r) gives flux density J'; R_s' is the corresponding resistivity (s m^{-1} MPa^{-1}). Now if A_r were to decrease, to sustain J either R_s' or ψ_x must fall. As mentioned earlier, there have been a number of experiments which indicate that R_s' does fall, as demonstrated by constant $\Delta\psi$ while J increases, but there appear to be no unequivocal explanations for the phenomenon. However, it seems unlikely that changes in R_s'—if these occur—can compensate for large reductions in A_r.

On the macroscale of root systems of plant communities in the field when the soil is wet, water uptake is most rapid where root systems are most dense. This is intuitively obvious because when L_V is high the pathlength for water movement from soil to root is short. However, in the absence of rain, the

7 Water Relations

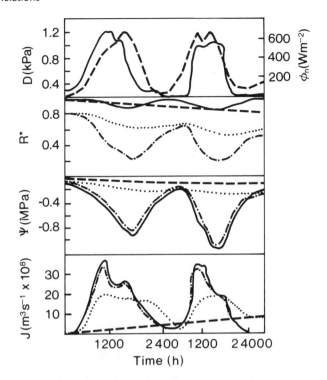

Fig. 7.6. Top: time courses of net radiation (φ_n, W m^{-2}, full curve) and vapour pressure deficit (D, kPa, broken curve) for a two-day period. The data were used to evaluate transpiration rates and hence changes in relative water content (R^*) and water potential (ψ, MPa) of leaves(———), branches ($-\cdot-\cdot$), stem top ($\cdots$), and stem base ($----$) of a *Pinus contorta* tree for which the appropriate capacitance data and information on change in K_c with θ were available. The bottom graph shows the simulated flows in the various segments (redrawn from Jarvis *et al*, 1981).

usual distribution of roots in the soil leads to rapid drying out of the surface layers where root concentrations are highest so that ψ_s—and hence $K(\psi)$ (equation (7.9))—falls and the difference ($\psi_s - \psi_x$) needed to extract water from those layers increases. The extraction zones then tend to move downwards to regions where the soil is wetter, although L_V in those regions may be lower. The extraction patterns shown in Fig. 7.7 are fairly typical (cf discussion on optimal root systems, Chapter 6).

Although the treatment so far has been in terms of single trees, the concepts discussed in this chapter can equally be applied to stands. In the case of stands the flux, J, through the average tree is (flux density per unit

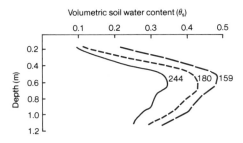

Fig. 7.7. Changes in soil moisture content between day of the year 159–244, measured by Hinckley and Bruckerhoff (1975) under a 55-year-old stand of oaks in Missouri. The data do not show a progressive drying pattern—there was rainfall during the measuring period—but do show fairly typical patterns of water extraction. Most of the roots were in the upper 0.5 m of soil. Pre-drawn xylem water potential (ψ_x) values were −0.14 MPa (day 159); −0.46 MPa (day 180) and −1.91 MPa (day 244).

land area)/(tree population per unit area). If, as is usually the case (see McNaughton and Jarvis, 1983; Whitehead *et al.*, 1984), transpiration from a forest is dominated by the aerodynamic term (see equation (3.51)), then increases in L^* as canopies develop will result in proportional increases in g_c (equation (3.46)) and increasing flow through individual trees. If transpiration were dominated by the energy term in equation (3.50), so that increases in L^* beyond $L^* \approx 3$ result in no further increase in absorbed energy, then the flow rate through individual trees will not change as L^* increases. The consequences of changing the populations in terms of exploitation of the soil by roots, and effectiveness of utilization of available soil moisture, are areas which require both experimental and theoretical research (see Landsberg and McMurtrie, 1984). Flow resistances through trees, relationships between sapwood cross-sectional area and foliage mass (Chapter 3) and sapwood cross-sectional area and flow resistances (R_x) are all areas where additional information is needed, although Whitehead *et al.* (1984) have recently provided valuable data for *Pinus sylvestris*. They found that, on plots with 608 and 3280 stems ha^{-1} (leaf area per tree = 39.5 and 9.5 m^2, respectively, $L^* = 2.4$ and 3.1, respectively) transpiration rates on the low population plot were about 0.7 times those on the high population plot, implying much higher fluxes through the trees on the plot with the lower population. Hydraulic resistance to flow through these trees was

7.5 × 10⁵ MPa s m⁻³ compared to 16 × 10⁵ MPa s m⁻³ for the trees on the high population plot.

7.5 THE HYDROLOGICAL BALANCE

We have seen that the water status of a particular tree at any time depends on the rate of water movement through the tree and the resistances in the water movement pathway. An important factor affecting these resistances is the amount of water in the root zone, and this depends on the hydrological balance, i.e. the balance between water input by rainfall and loss by run-off, drainage and evaporation and transpiration, integrated over any period of time.

In many countries important water supplying catchments are forested. With the increasing pressures on the world's fresh water supplies for direct human consumption, irrigation and industry, the need to improve our ability to maximize the amount of rainfall collected in catchments is becoming ever more urgent. As a consequence there have been a number of studies on the hydrology of forest catchments and comparisons of the water yield from forested and non-forested catchments (see review by Bosch and Hewlett, 1982). However, we are not concerned, here, with the water yield of forests, but with the water balance of a volume of soil supporting trees. The basic equation used for the analysis is the same—although the term of major interest is different.

If $z(m)$ is the depth of soil exploited by roots, the amount of water available to the plant community over time interal Δt is $(\theta_s(t+\Delta t) - \theta_m) z$, where θ_s is the volumetric water content of the soil and θ_m is the lower limit of available water. An estimate of θ_m is provided by the average value of θ_s at which average soil water potential in the root zone (ψ_s, see equation (7.7)) is -1.5 MPa—the so-called "wilting point". This is an old concept and only provides an approximation to the lower limit of available water, but in view of the errors involved there is little point in making too many complex calculations to try to estimate θ_m. We utilize the basic hydrologic (water balance) equation describing the conservation of water (precipitation):

$$P_T - E_T - q_D - q_R - \Delta\theta_s = 0 \qquad (7.29)$$

where P_T is rainfall amount in the time interval of concern, E_T is the amount of evapotranspiration, q_D the amount of water which drains out of the root zone, q_R is the amount of run-off and $\Delta\theta_s$ denotes the change in soil water content over the time interval.

This may be written

$$[\theta(t+\Delta t)-\theta_m]z = [\theta(t)-\theta_m]\,z + P_T - E_T - q_D - q_R \qquad (7.30)$$

which rearranges to

$$\theta(t+\Delta t)-\theta_m = [\theta(t)-\theta_m] + \left(\frac{P_T - E_T - q_D - q_R}{z}\right). \qquad (7.31)$$

We see that given a starting value for $\theta(t)$, knowledge of z and rainfall data we could, at least conceptually, calculate $\theta(t)$ for any subsequent time. To give a simple example, if $\theta(t)=0.25$, $z=1500$ mm, $P_T=11$ mm, $E_T=23$ mm, $0=q_D=q_R$ then $(\theta(t+\Delta t)z = 12.25$ mm, or $\theta(t+\Delta t) = 0.25 - 12/1500 = 0.242$.

The drainage and run-off terms in equations (7.30) and (7.31) will not be considered in any detail, although they may be of considerable importance in the water balance of a particular site. Run-off from a non-saturated soil occurs when the rate of water input (application) exceeds the infiltration capacity. A rough measure of infiltration capacity is given by the saturated, hydraulic conductivity of the surface layers. Run-off will always occur when water is applied to saturated soil, regardless of the application rate. Run-off from forests is usually relatively slow and is generally only significant when the upper layers are truly saturated. With litter on the ground and countless barriers to the movement of water across the surface the capacity of tree-covered areas to accept water is usually high. This is one of the reasons why water supplies from forested catchments are usually stable. Loss by drainage includes both vertical and lateral flow in the soil and is likely to be significant when the root zone is saturated. Such losses are difficult to measure and will not be discussed further.

Not all rainfall is effective in replacing soil water because of interception and evaporation of intercepted water. Furthermore, rainfall is redistributed by vegetation so that, although the hydrological equation may describe the average situation over relatively large, homogenous areas, it will not be accurate on the microscale. Soil heterogeneity, combined with variable vegetation, may result in considerable variation in soil moisture content over small areas (see later comment on the effects of stemflow; §7.5.1).

Between rain events, if the drainage term in equation (7.31) can be taken to be zero, the water balance of a site is determined by the rate of loss by evaporation from the soil surface and transpiration through plants. Evaporation from the soil surface will not be discussed further here; a thorough treatment can be found in Hillel (1980b).

7.5.1 Interception Losses

Effective precipitation (P^*)—the precipitation that contributes to soil water re-charge—is total precipitation less the amount (I) lost by interception and evaporation from foliage and branches.

$$P^* = P_T - I. \tag{7.32}$$

Doley (1981) lists the results of a great many empirical studies of interception losses in relation to rainfall amounts and discusses the considerable differences observed in various forest and plantation types. (I has been measured at anywhere between about 5 and 60% of P^*.) McNaughton and Jarvis (1983) also reviewed the available information and give $I = 10$–55% of total annual rainfall, depending on L^*, crown structure, species and climate.

Interception loss has often been described by linear empirical relationships (Gash, 1979) with storm size (total rain per storm). Such studies may provide useful estimates for rough "broad-scale" calculations, but they are of limited value for detailed hydrological work or studies on plant growth patterns and distribution. For detailed work it is necessary to analyse the process of interception in terms of mechanistic models. The best available is that originally derived by Rutter *et al.* (1971, 1975) and recently simplified by Gash (1979). E. M. O'Loughlin (personal communication) has refined this to a readily computerized form utilizing the solution to the differential equation describing the rate of exchange of water stored on the canopy (C_s, mm depth);

$$\frac{dC_s}{dt} = (1 - f_p)\frac{dP_T}{dt} - E_1(C_s). \tag{7.33}$$

f_p is the proportion of rain that passes through the canopy without touching it and E_1 is the instantaneous rate of evaporation of water from the canopy surfaces. E_1 is assumed to be proportional to the wetted surface area, and hence to C_s.

If C_s^* is the canopy storage capacity when saturated (so $C_s \leq C_s^*$), and $\bar{E}_1$ is the average (assumed constant) rate of evaporation from the canopy during a rain event then, with $E_1 = \bar{E}_1 (C_s/C_s^*)$, equation (7.33) becomes

$$\frac{dC_s}{dt} = (1 - f_p)\frac{dP_T}{dt} - C_s(t)\frac{\bar{E}_1}{C_s^*} \tag{7.34}$$

If the canopy parameters C_s^* and f_p are known (they can be obtained from conventional in-canopy measurements of throughfall, stemflow and above-canopy precipitation) equation (7.34) can be solved analytically; the solution provides the algorithm for a computer program for which rainfall is the driving variable and canopy parameters and evaporation rate are user-definable. O'Loughlin estimated values of E_1 by comparing observed values of P^* with values calculated using a range of values of E_I. He found a value of 0.25 mm h^{-1} to be suitable for a eucalypt canopy. This value is typical of a wide range of coniferous forest canopies (see McNaughton and Jarvis (1983) for a list derived from the literature).

For very dense canopies f_p may be effectively zero, so that $P^* = 0$ until C_s approaches C_s^*. In most cases some water will reach the ground after the start of a rain event before canopy saturation, because of drip and (possibly) stemflow. The amount will depend on the nature of the canopy and of the precipitation. The direct throughfall component of light, fine rain is likely to be smaller—perhaps much smaller—than that of heavy rain where most of the water is contained in large drops. Furthermore, light rain will wet canopy surfaces more evenly, and in low wind speeds there will be less mechanical displacement of drops and hence—at least initially—less drip than in heavy rain. When raindrop sizes are large rainsplash on rigid surfaces (e.g. branches or leathery leaves) contributes to immediate throughfall.

Interception losses will be much bigger if rainfall is intermittent than if it falls continuously for relatively long periods. This can be illustrated by considering the hypothetical situation where each of a series of rainfall events is of order C_s^* mm, while the interval between events is long enough for all the water stored on the canopy to evaporate. In this case P^* could be near zero.

The rate of evaporation of water from wet canopies depends mainly on the atmospheric vapour pressure deficit (D), and on wind speed. The effectiveness of wind depends on the aerodynamic properties of the canopy—its height and roughness, which is characterized by the separation of the dominant trees, and the density of the foliage (see Chapter 3). Evaporation rates from wet canopies are roughly the same as transpiration rates (McNaughton and Jarvis, 1983). Evaporation of intercepted water from multi-layered canopies will not be larger (on a land area basis) than from single-layered ones. Evaporation from the upper layers will be comparable but, because of the high humidity and low wind speeds in the understory (see "Canopy Microclimate", Chapter 3), evaporation from lower layers may be negligible.

The above considerations indicate why water losses by canopy interception vary greatly in different situations and also indicate that, in plant ecological work, detailed study will be necessary to characterize water balance of particular habitats.

7.5.2 Redistribution of Rainfall

The distribution of the unintercepted component of rainfall obviously depends on canopy homogeneity. Water intercepted by the canopy and not evaporated either coalesces into large droplets which drip off a multitude of low points (leaf tips, twigs, etc) in the canopy, or it reaches the ground as stemflow. Drip may or may not be evenly distributed, depending on canopy type. Stemflow will inevitably cause gross distortion of the soil wetting patterns which would have occurred from uninterrupted rainfall.

Stemflow may begin early in a rain event but does not commence until the streaks of water collecting on the trunks of trees from the lower side of branches are established as continuous flow lines. Doley (1981) gives a table of data characterizing the partitioning of precipitation of tropical forests and woodlands. Stemflow has been observed to vary between zero and 39% of the total rain in a storm. As in the case of interception losses, such data are of limited value as a guide to what might be observed in any particular situation because stemflow amounts depend so strongly on the physical characteristics of canopies and on rainfall characteristics. However, many of the studies cited by Doley suggest that stemflow may be treated as a constant fraction of total precipitation; a mean value of about 1% would be representative for many subtropical forest types. This is not inconsistent with the constant value (stemflow = 1.6% of total precipitation) found by Gash and Morton (1978) in a careful study of a number of rain events in a Scots pine forest.

Herwitz (1982), who studied the redistribution of rainfall by trees in a tropical rainforest in Australia, noted that the sheltered undersides of branches represented detention storage that usually requires a high intensity rainfall event (>100 mm day^{-1}), or several consecutive days of substantial rain, before a thoroughly wetted condition is achieved and substantial stemflow commences.

Herwitz (1982) considered stemflow volumes in relation to the basal area of trunks, not the crown area. He argued that the volume of water expected at the base of a tree is that which would have been collected by a rain gauge occupying the same area as the trunk, therefore the volume of water actually delivered by stemflow should be expressed as a ratio of the expected volume. This is the "funnelling ratio". Some of Herwitz's data are presented in Table 7.1, which includes an assessment of the minimum infiltration area—the minimum area over which stemflow has to spread in order to infiltrate the soil. This was calculated by dividing the rate of water input from stemflow by the mean saturated hydraulic conductivity of the surface 0.2 m. The area is minimum because no allowance was made for throughfall.

Whether Herwitz's findings would be repeated in other forest types may be questionable, but they demonstrate that the distribution of water resulting from stemflow may have important effects on the water balance of trees, particularly as they enter a dry period. The amount of rain intercepted by the dominant trees of a canopy is likely to be proportional to their canopy area.

If rainfall is heavy enough to cause significant stemflow the soil at the base

Table 7.1. Redistribution of high intensity rainfall by tropical trees. (Data from Herwitz, 1982.)

Species	Trunk basal area (m^2)	Expected stemflow rate (m^3 s^{-1}×10^6)	Minimum infiltration area (m^2)[a]	Funnelling ratios[b]
Balanops australiana	0.061	118	2.17	156
Ceratopotalum virchowii	0.049	94	1.15	112
Cardwellia sublimis	0.127	24.6	0.33	12
Elaeocarpus sp.	0.159	30.9	0.27	10

[a] Stemflow volume/mean saturated hydraulic conductivity to 0.2 m.
[b] Volume of rainwater which would have been collected in a rain gauge occupying the same area as the trunk.

of the dominant trees of a canopy may be considerably wetter than that around the base of subdominants. The water available in the root zone of the subdominants, and understory vegetation, will then depend on their location and root distribution relative to the dominants. A compensating factor for these trees may, however, lie in the probability that they will experience reduced evaporative demand because of shading by dominant trees, wind speed reduction and high humidities below canopies.

7.6 Consequences of Water Stress

It will be clear from section 7.2 that defining "stress" is not simple. There has been much discussion about this at various times, some of it concerned with the definition of stress by analogy with the engineering concept. However, it is clear that there is no cut-off point, in terms of some measure of tissue water status, on one side of which plants are stressed, and on the other unstressed. Rather than attempting to define stress we should, perhaps, be more concerned to explain the effects of tissue water status on growth. This leads to the question of which water relations parameter should be used to define status and to analyse growth, i.e. which parameter should be used as the independent variable in any attempt to relate growth to plant water status. The answer to this question is clearly strongly dependent upon the growth process under study (see Hsiao, 1973).

Turgor pressure (ψ_p) is probably the most important factor, affecting processes such as cell division and elongation, and hence shoot and root extension. It is also the operational factor in stomatal opening and closing. One of the mechanisms by which plants adjust to water deficits, over periods of days or weeks, is turgor maintenance by osmotic adjustment (reviewed by Turner and Jones, 1980). This is easily understood by reference to equation (7.1); if ψ_π is reduced by high cell solute concentrations water will tend to move into the cell(s), so that ψ_p will remain high.

Pre-dawn water potential, measured regularly, may be used to provide a measure of water stress over periods such as a season, or of the differences in "effective" plant water potential at different sites. Its use in this way is illustrated by Collatz et al. (1976) and by Parker et al. (1982).

It is clear from equation (7.13) that the foliage water potential of trees on a particular site and at a particular time depends on the diurnal course of transpiration and water movement through the plants. The resultant drop in leaf water potential is relieved as transpiration rates fall later in the day. If soil is dry and average ψ_s in the root zone is low, the pre-dawn water potential will be relatively low and midday values of ψ_f will be low, particularly if the evaporative demand is high. Conversely, even if the soil is dry, ψ_f may remain relatively high on cool damp days.

The consequences of water stress may vary in relation to the length of time spent at low potentials and the severity of the stress. In the short term (hours) one of the best documented consequences of stress is stomatal closure as a result of loss of turgor in leaf cells; this leads to reductions in photosynthesis (see Chapter 5) and hence reduced growth. Other relatively short-term results (hours–days) may be changes in the concentrations of growth regulators such as abscisic acid and in cambial and shoot growth.

Some of the effects of water stress over longer periods are illustrated by an experiment reported by Larsson and Bengtson (1980). They used a plastic cover to exclude water from the roots of a Scots pine throughout a summer growing season in Sweden. Photosynthetic rates in the stressed tree declined through the season, but recovered fully within a week of re-watering. Cambial activity—reflected in the growth rings—was affected in the year of treatment and for two years afterwards. A year after the stress there were effects on shoot growth (reduced) but needle growth was not affected either in the stress year or in the following year. It is not possible, over such long periods, to identify the component of water potential responsible for the stress and its effects.

Bradford and Hsiao (1982), in their review of physiological responses of plants to moderate stress, noted that "the first sign of (longer-term) water stress is usually a restriction in foliage growth". Using data from crop plants they developed a simple analysis demonstrating a strong feedback effect— reduction in foliage results in less photosynthesis and assimilation, hence less foliage production and so on. Of course, reductions in foliage area also result in reduction in transpiration.

Severe or prolonged stress not only results in reduction in leaf emergence and foliage expansion, but also causes leaf senescence. Borchert (1980) and Reich and Borchert (1982) demonstrated that leaf shedding, and the consequent improvement in the water balance of some tropical trees, triggered flowering. Thus water stress affects phenology.

Brix (1972) found that irrigation caused an increase in leaf size in Douglas fir—and hence leaf area. In the year following irrigation leaf size was not different between irrigated, fertilized, and irrigated + N-fertilizer treatments, but leaf number was greatly increased by treatments. Stem growth was increased by irrigation + fertilizer in the treatment year and somewhat less by irrigation alone. There was a strong irrigation × fertilizer interaction—the response to the combined factors was about twice as much as the sum of the responses to each separately. Brix concluded that water stress had a more adverse effect on cambial growth than on photosynthesis.

Many agricultural water use/irrigation/water stress experiments are assessed in terms of water use efficiency (WUE)—the amount of dry matter produced per unit water transpired. There are no data on this for trees but it may be a useful indicator of responses to water stress and the ability of plants to produce in spite of it. Preliminary estimates of water use efficiency could be made from evapotranspiration calculations and total dry matter production estimates. The calculations should include litter fall, tree mortality and other losses; WUE in terms of, say, harvestable stems will obviously be very much lower than for total dry matter production. Differences between genotypes may be another valuable indicator of the ability of trees to grow well when short of water.

The effects of stress on trees will vary depending on whether the stresses are applied suddenly or develop gradually. Trees growing in well watered regions subjected to sudden drought are likely to suffer more severely than trees acclimated to drought. (Acclimation describes the physical and physiological changes that occur during the life of a plant as a consequence of drought. Adaptation reflects the genetic or evolved properties or characteristics which make a particular species better able to tolerate drought or cold, heat or other stresses.) The sudden onset of drought may result in wilting and leaf shedding. In trees acclimated to drought osmotic adjustment may occur, resulting in turgor maintenance and higher leaf conductance at low ψ_f values. Sustained periods of water stress may also lead to a shift in root: shoot ratios in favour of roots—in effect, an increase in the carbohydrate partitioning coefficient for roots (see equation (5.1)).

Adaptation to drought is likely to be reflected in properties such as specialized leaves or leaf properties, phenological responses and inherent root characteristics. Leaves may be narrow, with high reflectivity, so that they are efficient heat exchangers (see the discussion on leaf energy balance, Chapter 3). The leaves of plants adapted to drought are often relatively rigid with waxy surfaces. Some, such as many of the eucalyptus species, orient their leaves away from the sun to reduce absorbed energy. Phenological responses include the capacity to produce flowers when drought is broken. Some trees (again eucalyptus species provide examples) have very large, deep root systems, enabling them to exploit very effectively water stored deep in the soil. That this is an adaptive trait is indicated both by empirical evidence (e.g. Carbon *et al.*, 1980) and by the evidence that structural differences in tree root systems are highly heritable (Nambiar *et al.*, 1982).

7.7 CONCLUDING REMARKS

In my view the final objective of all studies in plant–water relations is to arrive at the ability to predict tissue water status and, from functional relationships between the appropriate parameter and growth, evaluate the effects of water status on growth. We should be able to specify soil type, canopy structure, leaf area and stomatal responses, and weather conditions, and from this information calculate the time course of water status over any period of interest. Soil type would be specified in terms of soil water-holding characteristics (equation (7.7)) and probable rooting depth. The first step, using the physical models in Chapter 3, is to calculate the soil water balance (equations (7.30) and (7.31)) and the average water potential in the root zone. Information on the resistances in the flow pathways is needed so that the equations presented in this chapter can be used to calculate the plant water potential. In principle, all this can be done now, but besides the uncertainties in the values of many of the parameters required (root system resistances to flow, stem resistances), and the lack of any values at all for most tree species, there are two major obstacles in the way of establishing functional relationships between the water status of trees, and their growth patterns. The first is the variation in water potential from one part of a tree to another. The second is the problem of averaging water status in time.

That there must be spatial variation in ψ_f is obvious. Only in very open-structured trees, on days of largely diffuse radiation, will the energy load on all leaves—and hence their rates of transpiration (see equation (3.31))—be approximately equal. Not only are leaves at the top of canopies (or on different sides of isolated trees) normally subject to different energy loads, but the pathlengths along which water must move to different parts of the tree are different—and possibly of different characteristics (see Richter, 1973, for an analysis, and Hellkvist *et al.*, 1974, for illustrative data). The question therefore arises: how do we average tissue water status to produce a value which is useful as a basis for analysing growth? There is no straightforward answer to this, and probably little profit in speculative discussion, but the problem must be recognized, and deserves attention.

The problem of averaging water status in time arises from the diurnal pattern of water potential (Fig. 7.3), driven by transpiration. If some "critical" value of any particular parameter could be specified, above which a process (or processes) was not affected by tissue water status, and below which, say, there was a linear decline in the rate of the process down to some point at which the rate reached zero, then the analysis would be simple. The relationship proposed to account for the effects of water potential on stomatal conductance (Fig. 4.6, equation (4.7)) is almost that simple, although the extent to which it reflects reality is perhaps arguable.

The situation with regard to processes such as cell expansion and hence stem elongation is much less simple. The first obvious assumption—that such processes are likely to be related to cell turgor, and hence that we can analyse them in terms of average ψ_p, over appropriate periods—may be less tenable than it at first appears. This is because, in the dynamic daytime situation where transpiration rates are normally changing all the time, turgor pressure cannot be taken to be constant. Furthermore, even if the tree can be regarded as being in steady state with regard to water flows, changes in cell volume would cause changes in turgor in particular cells or masses of tissue.

To establish longer-term relationships between, say, dry matter production and soil water balance (equations (7.30) and (7.31)) there appears no alternative, at the moment, to sheer empiricism at the level of the stand. Over periods such as weeks, growth is likely to be a simple function of intercepted radiant energy (see Chapter 8) and water relations will be one of several factors that modify the effectiveness of energy conversion. To study such relationships, measurements of the soil water balance and of pre-dawn tissue water potential should be made in association with appropriate growth measurements. There have been few systematic attempts to define the effects of water stress on trees in terms of growth patterns, water use and measured tree water status over a long period. Such studies are badly needed. They should, preferably, also include factors like fertilization, while the use of irrigation, as in the work reported by Aronsson and Elowson (1980), is an invaluable component of such research.

When we achieve the objective of being able to calculate (accurately enough) the water status of trees on a particular site at a particular time, and can use that information to evaluate the effects of water status on growth patterns and productivity, we will have a useful tool that can be used in site classification, and estimates of yield potential, especially in countries with regions where water severely limits forest growth. No current site classification system includes water balance calculations.

7 Water Relations

APPENDIX TO CHAPTER 7

Worked example. Simulation of the time course of leaf water potential (ψ_f) and flows of water between xylem and storage tissue.

The procedure, using (7.26) and (7.27) is explained in complete detail. The processes being simulated are in fact continuous, but the procedure used here involves, in effect, removing a given volume of water ($J_f \Delta t$) from the leaves, updating ψ_f, allowing water to move through the xylem in response to the gradients so created, and so on.

Required. Starting values for ψ_s (say 0.1 MPa), ψ_f (0.1 MPa) and ψ_c (0.1 MPa); values for the resistances to flow from soil to xylem conducting tissue of roots and stem (R_s, taken as 3.5×10^6 MPa s m^{-3} for volume flux of water), from xylem to leaf (R_f, also taken as 3.5×10^6 MPa s m^{-3}) and from storage tissue (sapwood) to xylem ($R_c = 9 \times 10^6$ MPa s m^{-3}). ψ_s would probably be taken as constant over periods up to a day. Also required: functional relationships between volumetric leaf water content (θ_f) and ψ_f, and between volumetric storage tissue water content (θ_c) and ψ_c.

Equations (A.7.1) and (A.7.2) were derived for apple leaves; they are written in terms of volumetric water content but could as well be written in terms of R^*. The important requirement is that it must be possible to calculate the mass (hence volume, v_f^*) of water in the leaves at full turgor. Equations (A.7.3) and (A.7.4) were derived from data in the literature. They provide results consistent with equations (7.14) to (7.16) and the relationship between the relative water content of wood, and ψ, given by Waring *et al.* (1979).

$$\theta_f = 0.6 \, |\psi_f|^{-0.05} \, (\text{m}^3 \text{ water/m}^3 \text{ leaf}) \qquad (A.7.1)$$

$$\psi_f = 3.66 \times 10^{-5} \, \theta_f^{-20} \, (\text{MPa}) \qquad (A.7.2)$$

$$\theta_c = 0.4 \, |\psi_c|^{-0.25} \, (\text{m}^3 \text{ water/m}^3 \text{ wood}) \qquad (A.7.3)$$

$$\psi_c = 0.025 \, \theta_c^{-4} \, (\text{MPa}). \qquad (A.7.4)$$

From equations (A.7.1) and (A.7.3) the initial values of θ_f and θ_c are 0.67 and 0.71.

Tree Properties

Assume a community of trees with population $p = 500$ ha$^{-1} = 0.05$ m^{-2}. If leaf area index (L^*) = 3, average foliage area per tree (A_f) = 60 m^{-2}. Assume fresh leaf mass per unit area $\sigma_f = 0.3$ kg m^{-2}, hence leaf mass = $0.3 \times 60 = 18$ kg. We also need to know leaf water content at full turgor; say $w_s^0/W_f = 0.7$, hence weight of water in 18 kg leaf = 12.6 kg, i.e. 12.6×10^{-3} m^3.

It is necessary to have an estimate of the volume of wood storage tissue, V_c. For the purposes of this example let us assume that standing wood volume is 300 m^3, i.e. 0.6 m^3 tree^{-1}; of this 0.2 m^3 is sapwood which acts as storage tissue for water (ignoring roots and branches). The volume of water in store (v_c^* is $V_c\theta_c$, hence $v_c^* = 140 \times 10^{-3}$ m^3 tree^{-1}.

Transpiration Rate (E_t); Flow through Trees (J_f)

Transpiration rates (kg m^{-2} ground area) can be calculated from equation (3.51). Values of canopy transpiration rates for forests are given (for example) by Jarvis (1981). We will take E_t as constant = 0.25 mm h^{-1} = 0.25 kg m^{-2} h^{-1}, hence $J_f = 5$ kg tree^{-1} h^{-1} = 1.4×10^{-3} kg s^{-1} or 1.4×10^{-6} m^3 s^{-1}. If desired E_t can, of course, be varied for each time step.

Time step (Δt). The time step for simulation must be commensurate with the shortest time constant (RC) of the system. If Δt is too large, $J_f \Delta t$ will be too large, the computed change in ψ_f will be unrealistic and the model will be unstable and will cycle wildly. For example, if we take $\Delta t = 0.5$ h = 1800 s, $J_f \Delta t = 2.5$ kg = 2.5×10^{-3} m^3. Removing this amount of water from the leaves, v_f^* becomes $(12.6 - 2.5)10^{-3} = 10.1 \times 10^{-3}$ m^3, i.e. $\theta_f = 10.1/18 = 0.56$ and $R^* = 10/12.6 = 0.8$. Inserting this in (A.7.2) gives $\psi_f = -4$ MPa, which is unrealistic. Clearly, Δt must be much less than 0.5 h. We will assume $\Delta t = 0.083$ h = 300 s.

Procedure

1st cycle. (1) Inserting the starting values in equation (7.26), the first term on the right-hand side is zero, hence

$$V_f \Delta \theta_f = -J_f \Delta t = \Delta v_f^* = -1.4 \times 10^{-3} \times 300 \text{ m}^3.$$

To obtain $\Delta \theta_f$, V_f can be eliminated as follows

$$\theta_f = u_f^*/V_f$$

hence

$$\theta_f(t + \Delta t) = \theta_f(t) + \Delta \theta_f = \frac{v_f^*(t) + \Delta v_f^*}{V_f}$$

7 Water Relations

$$= \frac{(12.6) - (1.4 \times 10^{-3})300}{V_f} = 12.18.$$

The initial value of θ_f was 0.67, i.e.

$$\frac{12.6 \times 10^{-3}}{V_f} = 0.67$$

therefore $\theta_f(t + \Delta t) = (12.18/12.6)0.67 = 0.65$.

(2) Insert the starting values in equation (7.27):

$$V_c \Delta \theta_c = \left[\frac{1}{9 \times 10^6} \left(\frac{3.5 \times 10^6(-0.1) + 9 \times 10^6 \times (-0.1)}{(3.5+9) \times 10^6} \right) - \frac{0.1}{9 \times 10^6} \right] 300$$

$$= -6.67 \times 10^{-6} \, m^3.$$

This is the flow out of sapwood storage (v_c^*).

$$\theta_c(t + \Delta t) = \frac{v_c^* + \Delta v_c^*}{V_c} = \frac{0.14 - 6.67 \times 10^{-6}}{0.2} = 0.71.$$

2nd cycle. Update value of ψ_f (equation (A.7.2))

$$\psi_f = 3.66 \times 10^{-5} \times 0.65^{-20} = 0.2 \, MPa.$$

In (7.26)

$$\Delta v^* = 300 \frac{(-0.1 + 0.2)}{7 \times 10^{-6}} - 14 \times 10^{-6} \times 300$$

$$= 4.29 \times 10^{-6} - 420 \times 10^{-6} = 416 \times 10^{-6} \, m^{-3}$$

$$= (\text{Input to leaves}) - (\text{Loss}).$$

The next value of θ_f would be calculated as before.
In (7.27)

$$\Delta v_c^* = \left[\frac{1}{9 \times 10^6} \left(\frac{3.5 \times 10^6 \times (-0.2) + 9 \times 10^6(-0.1)}{12.5 \times 10^6} \right) - \frac{0.1}{9 \times 10^6} \right] 300$$

$$= 7.6 \times 10^{-6} \, m^3$$

and so on.

Obviously such calculations would normally be done by computer, when they can be done for any required period. For calculations over a day transpiration rate would be allowed to vary realistically.

Various points arise: for example, the calculations as made here assume that water is lost uniformly from all leaves and that there is good root–soil contact between roots and soil. The model can be used to explore the consequences of various assumptions and parameter values, for example, changing values and ratios of the resistances to flow, and various parts can be tested experimentally. The values used for any particular community of trees should be chosen to suit those trees.

8 Synthesis

The stated aims of this book are to provide a framework of knowledge and understanding about the processes involved in forest growth, and the basic information needed to calculate the growth rates of forests and their responses to disturbance or stimuli.

Management consists essentially of taking some action that disturbs the steady state, or the direction of change, of a system—in our case a forest. The manager has to guess at, and to gamble on, the consequences of his actions, so any information or tool that helps him reduce the uncertainty must be of great value. Knowledge and understanding of the system are of qualitative value; the commercial forest manager will also require quantitative models of forest growth, productivity and response to change or disturbance.

There has been a great deal of work done on the development of various kinds of forest model. The yields of wood to be on expected sites of different "quality" with varying stocking (tree population density, p) are estimated from yield tables. In rate models, stand growth rate (usually in terms of m^3 wood ha^{-1} yr^{-1}) is expressed as a function of age, site index and stocking. Many such models provide no information about the type and relative sizes of the trees in a stand. However, stochastic models that include survival (or mortality) functions and tree size distributions provide information about the proportion of the population in particular size classes at any time. Survival functions may also be used to evaluate the consequences of thinning, which will change the size distributions. Recently we have seen the development of individual tree models of various types, the growth of the trees being assumed to be dependent on distances between individuals and a range of competition indices. The trees are then combined to form stands.

These forest models are of great value to the manager and, in many cases, of considerable statistical sophistication. They may also incorporate hypotheses about response mechanisms (e.g. competition), but in the end they must be derived from measurements of the size of trees at various ages, growing at specified stocking on sites of defined "quality". There is no universal definition of site quality so such models are "non-transportable" with the result that the analyses have to be re-done on data obtained in a particular area before the model(s) can be applied in that area. The whole procedure is somewhat circular—before growth can be predicted trees have to be grown and measured. Stand development models, which predict the time course of stand development, also suffer from the basic flaw (see Chapter 2) that they are expressed as a function of time—not some physically and physiologically sound driving variable. Therefore they cannot provide any information about the consequences of conditions or perturbations outside the "normal" range.

Given the criticisms in the previous paragraph, the question is whether more mechanistic, process-based models that utilize the information and relationships presented in this book, would be any better. I will attempt to demonstrate in this chapter that, for predicting forest growth rates and productivity, such models have much to offer. They should allow us to avoid the problems of circularity and site quality and the need to have trees growing on a site (or at least to have data from similar sites in the same area) before we can predict performance at that site. It will also be possible to calculate the growth curves of a much wider range of tree populations and evaluate the consequences of events such as drought, thinning or fertilization. However, process-based models can only simulate the growth of a stand, or of an individual tree with specified characteristics; they offer no substitute for survivorship curves or for tree size distributions although they can, in principle, be combined with such statistical models. We also have to recognize that, even in terms of yield prediction, process-based models have a long way to go before they are better than conventional empirical growth models; the argument is not that they are, but that they can be.

There are two "families" of mechanistic growth models. The first is the detailed physiological "bottom-up" type model, which synthesizes growth by calculating the actions and interactions of the physiological processes contributing to it. This is essentially a research tool. The second is the empirical "top-down" type model. Such models use some simplified formulation(s) of the main physiological processes contributing to growth and the way they respond to one or more major driving variables. Yield data, as the expression of these processes, are analysed in terms of the driving variable (e.g. radiant energy) and modifying factors. The empirical constants and coefficients so derived must be consistent with physiological responses, as determined by detailed studies. There will, initially, still be a degree of site specificity but this should reduce as the relationships become more general. It is these "top-down" models which will provide the alternatives to current conventional forest growth models. The next section outlines the general characteristics of both detailed physiological and "top-

8 Synthesis

down" models; in later sections of this chapter a "top-down" model is developed in some detail.

8.1 MODELS: THEIR CHARACTERISTICS, VALUE AND LIMITATIONS

In the brief comment in Chapter 1 a model was defined as "a formal and precise statement or set of statements embodying our current knowledge or hypotheses about the working of a particular system and its responses to stimuli". When such statements are made in mathematical terms it usually becomes clear that our knowledge is incomplete and assumptions have to be made about how parts of the system work. The consequences of these assumptions can be explored, either algebraically or numerically, and it must be possible to test them, and the model as a whole, experimentally. (By these criteria, conventional forestry models scarcely qualify as models; they are not hypotheses but descriptions of observations. However, this is a matter of semantics.)

8.1.1 Detailed Physiological Models

Models can be written at any level and many have already been presented in this book, e.g. equations (3.23)–(3.28) comprise a model of energy interception by canopies; equation (3.31) is a physical/physiological model of the energy balance of a leaf; equation (4.10) is an empirical model of leaf photosynthesis; equations (5.1), (5.4) and (5.5) are carbon partitioning models; equations (6.10) and (6.3) describe growth in relation to nutrition and the movements of nutrients within plants; equations (7.26) and (7.27) model water movement in trees. All these are process models and, at their level, they are all empirical approximations to reality; various parameters must be measured to provide values for the constants and coefficients needed for the equations and we must be clear about the degree to which we expect them to reflect reality. The idea of different levels of organization was discussed in Chapter 2. To recapitulate briefly: plant growth may be considered in terms of processes at a number of levels (see Fig. 2.2). Process-based models are explanatory, in the scientific sense of understanding how something works in terms of the processes involved at the next (less complex) level down. So we try to understand the growth of a forest in terms of the growth patterns of the trees in relation to their environment and their neighbours; we try to understand the growth of the trees in terms of carbon production and allocation and the factors affecting these processes, and so on. Process-based, explanatory models should be much more widely applicable—more general—than empirical models, at any level.

All the models mentioned in the previous paragraph could be the end point of sub-models at the next level down. For example, it was pointed out in Chapter 3 that equations (3.23)–(3.26) are simplified versions of Norman's (1982) more complex treatment; the r_s term in equation (3.31) is itself the subject of a great deal of modelling at the cellular levels (see Jarvis and Mansfield, 1981), and so on. However, for present purposes, we will take these process models as the lowest level of organization with which we are concerned.

We can build such process-based sub-models into a complete model of the growth of a tree or stand. Some elementary calculations of this type were done in Chapter 5, to give estimates of canopy photosynthesis; a reasonably complete model is outlined in Fig. 8.1. This illustrates a series of sub-models (each of the oval process "boxes") with response times of hours or days, that would be linked in a hierarchical system of sequential calculations, after each round of which the state of the plant would be updated. Not all the calculations need to be done with the same frequency. For example we have already seen (Chapter 5) that there is probably little point in calculating stem respiration over periods shorter than a day; uncertainties in stem temperature, if nothing else, would render spurious the apparent increase in precision derived from more frequent calculations. Similarly there is little point in updating soil water at intervals of less than a day. (The soil-water/root distribution sub-model could itself be of great complexity or relatively simple.)

Models such as that outlined in Fig. 8.1 allow exploration of the consequences of varying the rates of various processes, and of exploring the sensitivity of the system to changes, uncertainties and variations in parameter values. They provide a means of integrating, in quantitative form, knowledge about processes and, like all models, they provide a framework for research programmes. This matter of integration is important. Without the framework of models it is difficult to see what practical contribution is made by much reductionist research. ("Reductionist" means that problems are reduced to the simplest possible components, or low-level processes.) Such research is essential and often intellectually exciting, but we need some means of evaluating the significance of the results in relation to the functioning of the whole system. Relatively complex, process-based models, provide the means.

There are many critics of such models. Some scientists argue that they should not be developed because our knowledge of virtually all the processes involved is inadequate and many of the "constants", coefficient and parameter values will be inaccurate. It is also argued that, because of the

8 Synthesis

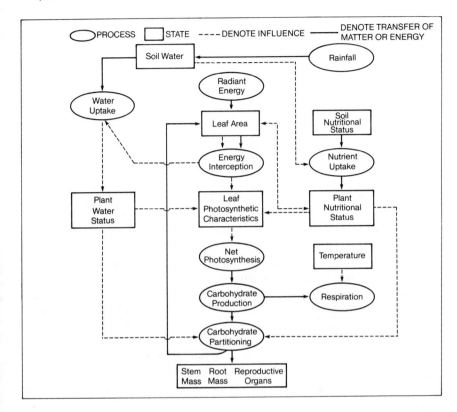

Fig. 8.1. Schematic representation of a detailed mechanistic model of tree growth. (It could, of course, be applied to virtually any plant.) Each of the physical/physiological processes (water uptake, energy interception, net photosynthesis, etc) would be represented by sub-models. The complexity of these sub-models would depend on the information available for the system under study—and the skills and propensities of the modeller. The time intervals used for the various sub-models need not necessarily be the same (see Landsberg, 1981b); the period for which the state of any component of the system can be assumed steady varies. Such a model would normally span two levels of organization.

empiricism of the process models, they are not universally applicable. These arguments have substance, but those who use them seldom seem to suggest an alternative to the use of such models. There is, and always will be, need for research at lower levels of organization, but we also need some method of integration, a framework within which to evaluate our knowledge at various levels and the consequences of our guesses about interactions. If we waited until we had "adequate" (perfect?) knowledge of even some of the processes in plant biology, we would wait a very long time, accumulating ever more information at ever lower levels of organization, but never venturing to evaluate the workings of the whole system for fear of being wrong. We will undoubtedly develop models which are wrong, but if these are linked to a rigorous experimental programme and always recognized for what they are—the best expression of our current knowledge—then we can only learn from our mistakes.

Complex models must not be used uncritically. There is always a danger that having developed such a model, and having it "successfully" run on a computer, there will be uncritical faith in the results. It is essential that the modeller be constantly aware, and make his associates aware, of the assumptions involved and the limits to the accuracy of the parameter values. Complex models should be constructed in such a way that, if possible, errors in the sub-models are not propagated through all parts of the model. Unfortunately, this may sometimes be unavoidable: for example, bad estimates of leaf nutritional status, affecting leaf area and photosynthetic characteristics, will affect carbon production and hence, via the partitioning sub-model, root production, soil exploration and so back to nutrient uptake and plant nutritional status. The example of leaf nutritional status is particularly pertinent because, as noted in Chapter 6, our ability to calculate nutrient uptake by trees, whatever information we are given about soil chemistry and water content, is poor. This is probably one of the most uncertain areas in this type of modelling. Each sub-model should therefore be written in such a way that it can be tested separately and the output clearly identified.

Ideally there will be constant feedback between modelling and experimental programmes (see 8.1.3). Experiments at the level of the sub-models test their accuracy and lead to constant extensions of their generality and improvement of the models. Tests of the end result are discussed briefly later.

8.1.2 "Top-down" Simple Models

"Top-down" models are so-called because they describe the responses of the system as a whole to changes in its main driving variables. They are, at least initially, likely to be simpler than detailed physiological models, and of much more immediate value to management. If they are to be adopted as management tools, the driving variable must be simple and readily available to managers.

It emerges from the preceding discussion that the ideal approach would be to develop both "top-down" and detailed models of the same system. "Top-down" models can be developed to the point of practical usefulness much more rapidly than "bottom-up" process-based models. The consequences of the various simplifications embodied in a "top-down" model can be evaluated using the detailed physiological model, run over periods (say weeks to months) which overlap with the response time of the simplified model. The value of the procedure will emerge from the later discussion of a "top-down" model (see also Landsberg, 1981b).

8.1.3 Model Testing

The definition of models as hypotheses implies that they should be regarded as statements about the working of systems—or parts of them—that cannot be taken as definitive, and must be tested. Models should therefore be written in forms which allow them to be tested. This is one of the reasons why mathematical models have great advantages; the statements are precise and it is generally possible to devise experiments which could invalidate the model. We should note that, in strict logic, no model can be proved—it can merely be "not invalidated". The accumulation of data and observations consistent with the predictions of models certainly give increased confidence in their use, but a single instance of disproof is sufficient to demonstrate that a particular model cannot be universally applied. These considerations are part of the philosophy of science, an area of great interest and, of course, of considerable debate. It is not appropriate to venture into it here, but it may be appropriate to suggest that all those who are concerned with modelling—which in my view should be all scientists[†]—should be reasonably clear about the underlying philosophy of, and reasons for, models.

Scientists should also be clear about model testing: the criteria for invalidating—and hence discarding—them, and the degree to which it is permissible to "prop up" hypotheses by subsidiary hypotheses to explain inconsistent data and observations, and so on. In the context of the physiology and behaviour of trees in forests—as in agriculture and ecology—testing may range from field observations to detailed physiological measurements in the laboratory. Testing is easier at lower levels of organization—which is not to suggest that the techniques involved at these levels may not be difficult and expensive, but the models are usually more precise. One can test the form of equations such as (3.31) and (4.10), or (7.26) and (7.27), by using well established physical techniques. The parameter values may, of course, vary between samples and almost certainly between species, but this may not invalidate the general statements embodied in the equations. However, testing the output of a canopy carbon balance model is a vastly more complex exercise since it involves evaluating the performance of a number of

---

[†]Although I am advocating, in the context of tree and forest physiology, the use of mathematical models, conceptual, qualitative models may be all that are needed in many fields.

sub-models (e.g. Fig. 8.1) in a highly variable environment. Such testing would involve weather (micro-climatic) measurements, growth measurements, leaf sampling and possibly stem boring to evaluate nutrient status and movement, photosynthesis and water relations measurements, soil and root studies. All the inputs and conditions have to be specified and the end result accurately evaluated, but the complexity and difficulty of this type of research does not provide a reason for not developing detailed models.

Another point worth mentioning is the difficulty of learning by experimentation how some systems work. For example, the complexity and variability of tropical rain forests make conventional experiments extremely difficult, added to which there may be reasons—such as environmental considerations—why it would be highly undesirable to disturb such systems.

"Top-down" models will usually be derived from biomass data, growth measurements and weather and soil data (see 8.2). Similar data will be required to test them, with the (obvious) proviso that models should not be tested against the data used to derive them. In the next section we consider the development of a "top-down" forest productivity model and how parameter values might be obtained for it.

8.2 A "TOP-DOWN" MODEL OF FOREST PRODUCTIVITY

The basic model is illustrated in Fig. 8.2. The justification for this form of model comes from agriculture (see Gallagher and Biscoe, 1978; Monteith, 1981), but Linder (1985) has recently demonstrated that the basic relationship (equations (8.1) and (8.2)) holds for trees.

Growth (dry matter accumulation) depends on the balance between production by photosynthesis and losses by respiration and—particularly important in forests—mortality and turnover of whole trees, as well as branches, leaves, roots and fruits of all individuals. Photosynthesis is driven by radiant energy so we may assume that dry matter production per unit area (ΔW) over time interval Δt is proportional to the amount of short-wave radiant energy absorbed by the canopy (φ_{abs}) over the same period, i.e.

$$\Delta W \propto \varphi_{abs} \Delta t. \tag{8.1}$$

If Δt is a short period (say a week), relative to the life span of the forest, equation (8.1) implies that the mass at time t ($W(t)$)—after n such periods—will be

$$W(t) = \varepsilon_\varphi \sum_{i=1}^{n} \varphi_{abs\,i} \Delta t. \tag{8.2}$$

8 Synthesis

ε_φ may be regarded as an energy conversion efficiency term and will have units of kg MJ^{-1}. Values for ε_φ will have to be derived from time-series biomass determinations, using equation (8.2). For closed canopies φ_{abs} would be calculated from equation (3.28); for incomplete canopies equation (3.29) could be used.

It may be that equation (8.1) should be written $W = f(\varphi_{abs})$ and that the functional relationship is non-linear. There are several possible reasons for variation: (i) φ_{abs} may increase because the length of the growing season is increased by leaf retention. This is unlikely to cause non-linearity. (ii) φ_{abs} may increase because L^* increases, or (iii) because φ_s increases. Both these cases could cause non-linearity. These possibilities should be investigated, but to simplify the development and discussion here we will assume that the linearity implied by equation (8.2) holds (as it does for cereal crops up to anthesis).

It seems highly unlikely that ε_φ will be constant for all sites and growing conditions, but it may be effectively constant for a single species growing under non-limiting conditions, i.e. given appropriate data sets we should be able to determine a single "potential" value of ε_φ for a species, reflecting the genetic potential of stands of that species.

To account for the effects of environmental limitations, which will reduce growth below its potential level, we introduce a series of modifiers. The main

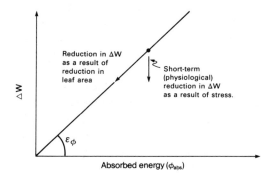

Fig. 8.2. The basic empirical relationship underlying a "top-down" model to calculate dry matter production (ΔW) over some time interval Δt during which an amount of radiant energy φ_{abs} is absorbed by the canopy. The slope of the relationship (ε_φ) is, in effect, an efficiency factor, specifying the efficiency of conversion of electromagnetic energy into chemical energy (equivalent). We would expect ε_φ to be affected by nutrition and temperature (both of which may reduce it) and by water. Over short periods the effects of water would be physiological; over longer periods we would expect it to cause reduction in the leaf area and hence in φ_{abs}. ΔW must be partitioned into the various components of interest.

environmental modifiers are water, temperature and nutrition; the way they (are assumed to) operate, their basis in physiology and the way they might be derived are described below. The mechanistic basis of each of these modifiers could—and should—be described in terms of the various sub-models already discussed. The modifiers must act on ε_φ. They do *not* account for effects on *growth* as such, but for effects on the efficiency with which absorbed radiation is utilized, i.e. they are primarily effects on photosynthesis, respiration and turnover processes.

8.2.1 Limiting Factors; Temperature

If we assume that the temperature for the processes contributing to growth was always in the optimum range (see Fig. 2.1) there would be no need to introduce a temperature modifier to equation (8.1). However, this is unlikely in most regions, so equation (8.1) will be written

$$\Delta W_i = \varepsilon_\varphi \varphi_{\text{abs } i} m_T^* \tag{8.3}$$

where m_T^* takes values between 0 and 1. Both ε_φ and m_T^* can be derived from measurements made where nutrients and water are (as far as can be guaranteed) non-limiting. The experiment in which *Pinus radiata* was irrigated with effluent by Cromer *et al.* (1982), probably provided such conditions. To determine m_T^* for the purpose of this model, where $\Delta t =$ one week, growth measurements would be made at weekly intervals and the average temperatures (T) over the same period used as the independent variable. Although the model is written in terms of total dry matter production, the function $f(T)$ (see Chapter 2) could be derived on the basis of measurements such as shoot and leader elongation made at periods of the year when incoming radiation does not vary too greatly. Alternatively, recourse should be had to multivariate analysis to separate the effects of T and φ_s. (This is consistent with the suggestion in 2.1 that "multivariate analysis is a technique that may be valuable in testing the relative importance of particular variables.") To derive m_T^* the growth rate in the optimum temperature range is taken as unity, and growth rates at temperatures above or below the optimum range are normalized to it; e.g. if leader growth rate at 27°C was 50 mm week^{-1}, and at 13°C it was 20 mm week^{-1}, the value of m_T^* at 13°C would be $20/50 = 0.4$. It is unlikely that any set of field experimental data would provide an adequate range of T, so the limit values and optimum range may have to be estimated from, for example, experiments done in controlled environments.

The procedure suggested above is simply aimed at obtaining a suitable empirical function to describe the dimensionless modifier m_T^*. In fact the

8 Synthesis

primary effect of temperature is probably on leaf area development, and hence on φ_{abs} and carbohydrate production. Temperature may also affect carbohydrate allocation and certainly affects enzyme functioning in cells. All these effects are contained in the temperature modifier. The physiological basis of m_T^* should therefore be investigated at these process levels, with a view to replacing it, when possible, with more detailed relationships describing the effects of temperature on carbohydrate production and allocation.

8.2.2 Limiting Factors: Nutrition

The nutrition modifier, again taking values between 0 and 1, is introduced into equation (8.1) in the same way as the temperature modifier, so we get

$$\Delta W_i = \varepsilon_\varphi \varphi_{abs\,i} m_T^* m_M^*. \tag{8.4}$$

Again we note that the primary effects of nutrition are on leaf area (see Chapter 6) and hence on φ_{abs} and carbohydrate production. This is how we would expect to describe the effects of nutrition in a process-based model (see Fig. 8.1).

However, as we have already discussed, the effects of site fertility on plant nutrient status and hence on growth are more difficult to model than the effects of either temperature or water. This applies at both the mechanistic and completely empirical levels. Fertilizer experiments would lead to estimates of m_M^* provided growth in these experiments was measured over intervals consistent with the model and during which water was not a limiting factor. Such data are extremely rare in forest research and must be gathered, for example from experiments involving irrigation and fertilization, short-interval growth measurements and careful monitoring of [M] in foliage and other tissues. Foliage nutrient concentrations [m_{fl}], widely measured in forest research, will provide guidance to uptake from different soil types although, because of their dynamic nature (see Chapter 6), they are not useful as the independent variables for analysing growth. In fact it is difficult to suggest a useful empirical method of arriving at values for m_M^*. The more mechanistic approach used by Ågren (1983, see Chapter 6) seems far more valuable and I would suggest that some version of this could serve as a basis for calculating the effects of nutrition on dry matter productivity. Ågren's model could be coupled to models of nutrient mineralization in forest soils.

8.2.3 Limiting Factors: Water

The primary effect of protracted periods of water stress on plants is to reduce leaf area (see Chapter 7). The water modifier, which like the others takes a value between 0 and 1, must therefore reflect this effect. In a mechanistic model dealing with short time periods (hours, days) water stress effects would probably be reflected through their effects on stomatal aperture (see Chapter 4), but effects on carbohydrate partitioning and hence foliage growth would have to be incorporated. For the purposes of the empirical "top-down" model, the water modifier is introduced as before, giving

$$\Delta W_i = \varepsilon_\varphi \varphi_{\text{abs}\,i} m_T^* m_M^* m_\theta^*. \tag{8.5}$$

m_θ^* is strictly a function of tissue water status but for a long-term "top-down" model can reasonably be made dependent on average soil water content in the root zone, θ_s. To calculate θ_s we use equations (7.30) and (7.31), with Δt being one week and drainage and run-off taken to be zero unless rainfall during the week exceeds E_T plus the storage capacity of the soil. In other words we calculate a weekly water balance. E_T could be calculated from equation (3.50), or more simply estimated: e.g. using an estimate of φ_n (equation (3.15)), and average Bowen ratio (equation (3.17)), possibly taken to be a function of rainfall—see Landsberg (1984).

m_θ^* can be taken as a linear function of θ_s, having a value of 1 where θ_s has values such that θ_s (see equation (7.8)) is not less than, say, -0.1 MPa, and zero when $\theta_s \approx -1.5$ MPa—the so-called wilting point.

8.2.4 Final Formulation and Use

The final form of the "top-down" model will therefore be

$$W(t) = m_M^* \varepsilon_\varphi \sum_{i=1}^n \varphi_{\text{abs}\,i} m_T^* m_\theta^* \Delta t. \tag{8.6}$$

m_M^* is outside the summation because it will usually be constant for any particular site. Note that the modifiers operate in a linear, multiplicative manner.

Alternative formulations should be investigated, such as

$$\Delta W_i = \varepsilon_\varphi \varphi_{\text{abs}\,i} c \frac{m_T^* m_\theta^* m_M^*}{m_M^* m_\theta^* + m_T^* m_\theta^* + m_T^* m_M^*} \tag{8.7}$$

which is a non-linear Michaelis–Menten-type response curve. Expressions similar to this have been found useful in a number of areas. If the constraint $\Sigma m_i^* = 1$ holds, the constant c would have a value of 3.

Inputs to the model are the tree population p, the values of ε_φ, m_T^*, m_M^*, and m_θ^*, weekly (or monthly, if that is the interval chosen), total φ_s for the site of interest, weekly (or monthly) values of E_T, rainfall (P_T) and temperature (T).

8 Synthesis

Information on soil–water characteristics and rooting zones will also be needed. The output (as envisaged here) is dry mass per unit area. After each time interval the dry mass produced must be partitioned, using relationships such as those presented in Chapter 5, and the leaf mass (and hence L^*) updated. For the period before full canopy, estimates would have to be made of the fraction of ground shaded by the trees (giving ξ_g in equation (3.29)); during this period the model will essentially simulate the growth of individual trees. The time of canopy closure will depend, to a large extent, on the population and on the proportion of dry matter allocated to foliage. It will also be necessary to introduce a foliage loss term (see equation (3.6)).

I mentioned earlier that this type of model should be linked with survivorship (mortality) curves of the sort used in conventional forest models. If, after full canopy, tree mortality did not affect canopy structure significantly, it would not affect ΔW, but since the average growth rate of individual trees is $\Delta W/p$, it would affect that. Similarly, thinning could be introduced (reduction in p) and would affect canopy structure by the corresponding reduction in L^*. Conventional modelling could be used to predict tree size distribution, provided that relationships between tree mass and volume are available.

The model can be run through a cycle of years and will give, at any time, the state of the stand. In its simplest form, i.e. having calculated φ_{abs} and m_T^* for each week, plus the water balance and hence m_θ^*, the calculations are trivial. However, at the level of the sub-models the calculations might be relatively simple or very complex. For example, estimates of m_M^* may be simple constants, or based on complex nutrient uptake modelling (probably not feasible at the moment); estimates of m_θ^* may be based on straightforward soil water balances and a linear relationship with θ_s, or complex tissue water status models (see Chapter 7).

The extent to which it is considered justifiable to use complex sub-models is a matter for the judgement of the user and the developer. It depends on the quality of the information available for the species and (geographical) region of interest, on how good the modifier sub-models and relationships between modifiers and growth are considered to be, and how well the particular version of the model has been, or can be, tested. Like any other model, "top-down" models should be constantly tested, revised and improved. As noted earlier, the ideal approach must be to develop simultaneously both "top-down" and "bottom-up" models, so that the empirical factors in the "top-down" model are gradually replaced by more mechanistic descriptions of the processes governing productivity. It may be that, for practical purposes, "top-down" models will be retained, but the implications and limitations of the simplifications involved in them should be well understood—and quantifiable.

8.3 CONCLUDING REMARKS

I noted, early in this chapter, that "the argument is not that, (physiologically based) models are better, than (conventional forestry models), but that, as estimates of stand productivity, they can be". My argument cannot be supported by data because, to my knowledge, no models of the types suggested here have yet been developed for forests. This is not to say there are no mechanistic forest models—there are a number, dealing with various aspects of forest dynamics (see, e.g. Hälldin, 1979). The US IBP programme (see Waring and Edmonds, 1974) resulted in a great effort in forest biological modelling but in my view the models produced were too complex to be useful, having a great many site-specific parameters. There are no "top-down" forest models. Hopefully the case for such models is persuasive enough for research scientists in forestry to look seriously at their possibilities: even relatively crude versions should have the potential for considerable flexibility, allowing exploration of the "what if" questions, such as those concerned with changes in ε_φ, changes in carbon partitioning, the consequences for growth of a wide range of thinning regimes, the consequences of drought and—in due course—the returns on fertilization.

The "top-down" model (equation (8.1)) fulfils the requirement that its parameters are soundly based in physiological terms; each can be derived in much more complex form from physical and physiological considerations. The model can also be "tested" by a detailed process-based model, such as that presented in Fig. 8.1. This will allow examination of the consequences of the assumptions involved in the simple modifiers, and may provide the means to improve them.

To conclude, the understanding and prediction of forest growth must be based on quantitative understanding of the physiological processes contributing to it, and on the responses of these processes to changes in their environmental conditions. Such changes may be abrupt or cyclic, man-made or natural, but most of them can be described in physical or chemical terms and, at least, conceptually, their consequences assessed. There is a very long way to go before forest scientists can do this with any surety for any kind of forest. I hope that, in future, more will be prepared to work towards the goal of real quantitative prediction, recognizing that although forests are complex biological systems they can usefully be analysed in terms of their component parts and processes, using mathematical, physical and chemical tools.

References

Ågren, G. I. (1983). Nitrogen productivity of some conifers. *Can. J. For. Res.* **13**, 494–500.
Albrektson, A. (1980). Relations between tree biomass fractions and conventional silvicultural measurements. *In* "Structure and Function of Northern Coniferous Forests—an Ecosystem Study". (T. Persson, ed.), *Ecol. Bull. (Stockholm)* **32**, 315–327.
Anderson, M. C. (1981). The geometry of leaf distribution in some south-eastern Australian forests. *Agric. Meteorol.* **25**, 195–205.
Aronsson, A., and Elowson, S. (1980). Effects of irrigation and fertilization on mineral nutrients in Scots pine needles. *In* "Structure and Function of Northern Coniferous Forests—an Ecosystem Study". (T. Persson, ed.), *Ecol. Bull. (Stockholm)* **32**, 219–228.
Atkinson, D., Bhat, K. K. S., Coutts, M. P., Mason, P. A., and Read, J. (1983). "Tree Root Systems and their Mycorrhizas". Martinus Nijhoff/Dr W. Junk Publishers, The Hague.
Attiwill, P. M. (1979). Nutrient cycling in a *Eucalyptus obliqua* (L'Herit.) forest. III. Growth, biomass, and net primary production. *Aust. J. Bot.* **27**, 439–458.
Attiwill, P. M. (1980). Nutrient cycling in a *Eucalyptus obliqua* (L'Herit.) forest. IV. Nutrient uptake and nutrient return. *Aust. J. Bot.* **28**, 199–222.
Attiwill, P. M. (1981). Energy, nutrient flow, and biomass. *In* "Australian Forest Nutrition Workshop: Productivity in Perpetuity", pp. 131–144. CSIRO, Melbourne.
Baker, T. G. and Attiwill, P. M. (1981). Nitrogen in Australian eucalypt forests. *In* "Australian Forest Nutrition Workshop: Productivity in Perpetuity", pp. 159–172. CSIRO, Melbourne.
Barnes, A. (1979). Vegetable plant relationships II. A quantitative hypothesis for shoot-storage root development. *Ann. Bot.* **43**, 487–499.
Batschelet, E. (1979). "Introduction to Mathematics for Life Scientists" Springer Verlag, Berlin.
Beadle, C. L., Turner, N. C., and Jarvis, P. G. (1978). Critical water potential for stomatal closure in *Sitka spruce*. *Physiol. Plant.* **43**, 160–165.
Beadle, C. L., Talbot, H., and Jarvis, P. G. (1982). Canopy structure and leaf area index in a mature Scots pine forest. *Forestry* **55**, 105–123.
Bell, C. J., and Rose, D. A. (1981). Light measurement and the terminology of flow. *Plant, Cell and Environ.* **4**, 89–96.
Benecke, U. (1980). Photosynthesis and transpiration of *Pinus radiata* D. Don under natural conditions in a forest stand. *Oecologia (Berl.)* **44**, 192–198.

Benecke, U., and Nordmeyer, A. H. (1982). Carbon uptake and allocation by *Nothofagus solandri* var. diffortioides (Hook, J.) Poole and *Pinus contorta* Douglas at montane and subalpine altitudes. *In* "Carbon Uptake and Allocation in Subalpine Ecosystems as a Key to Management". (R. H. Waring, ed.), pp. 9–21. Forest Research Laboratory, Oregon State University.

Björkman, O., Boynton, J., and Berry, J. (1976). Comparison of the heat stability of photosynthesis, chloroplast membrane reactions, photosynthetic enzymes and soluble protein in leaves of heat-adapted and cold-adapted C4 species. *Carnegie Inst. Washington Year Book* **75**, 400–407.

Böhm, W. (1979). "Methods of Studying Root Systems". Springer Verlag, Berlin.

Bowen, G. D. (1985). Micro-organisms and tree growth. *In* "Research for Forest Management". (J. J. Landsberg and W. Parsons, eds.). pp. 180–201. CSIRO, Melbourne.

Borchert, R. (1980). Phenology and ecophysiology of a tropical tree, *Erythrina poeppigiana* D. F. Cook. *Ecology* **61**, 1065–1071.

Borough, C. J., Incoll, W. D., May, J. R., and Bird, T. (1978). Yield statistics. *In* "Eucalypts for Wood Production" (W. E. Hillis and A. G. Brown, eds.), pp. 201–225. CSIRO, Australia.

Bosch, J. M. and Hewlett, J. D. (1982). A review of catchment experiments to determine the effect of vegetation changes on water yield and evapotranspiration. *J. Hydrol.* **55**, 3–23.

Bradford, K. J., and Hsiao, T. C. (1982). Physiological responses to moderate water stress. *In* "Encyclopaedia of Plant Physiology, Vol. 12B. Physiological Plant Ecology II: Water Relations and Carbon Assimilation" (O. L. Lange, P. S. Nobel, C. B. Osmond, H. Ziegler, eds.), pp. 263–324. Springer Verlag, Berlin.

Brix, H. (1972). Nitrogen fertilization and water effects on photosynthesis and earlywood-latewood production in Douglas-fir. *Can. J. For. Res.* **2**, 467–478.

Brunes, L., Öquist, L. and Eliasson, L. (1980). On the reason for the different photosynthetic rates of seedlings of *Pinus sylvestris* and *Betula verrucosa*. *Plant Physiol.* **66**, 940–944.

Burrows, F. J., and Milthorpe, F. L. (1976). Stomatal conductance in the control of gas exchange. *In* "Water deficits and Plant Growth" Vol. IV. (T. T. Kozlowski, ed.), pp. 103–152. Academic Press, New York.

Butler, D. R., and Landsberg, J. J. (1981). Respiration rates of apple trees, estimated by CO_2-efflux measurements. *Plant, Cell and Environ.* **4**, 153–159.

Calder, I. R. (1977). A model of transpiration and interception loss from a spruce forest in Plynlimon, Central Wales. *J. Hydrol.* **33**, 247–265.

Calder, I. R. (1978). Transpiration observations from a spruce forest and comparisons with predictions from an evaporation model. *J. Hydrol.* **38**, 33–47.

Campbell, G. S. (1977). "An Introduction to Environmental Biophysics" Springer Verlag, New York.

Campbell, G. S. (1981). Fundamentals of radiation and temperature relations. *In* "Encyclopaedia of Plant Physiology, Vol. 12 A. Physiological Plant Ecology I. Responses to the Physical Environment" (O. L. Lange, P. S. Nobel, C. B. Osmond and H. Ziegler, ed.), pp. 11–40. Springer Verlag, Berlin.

Cannell, M. G. R. (1982). "World Forest Biomass and Primary Production Data" Academic Press, London.

Cannell, M. G. R. and Smith, R. I. (1983). Thermal time, chill days and prediction of budburst in *Picea sitchensis*. *J. Appl. Ecol.* **20**, 951–963.
Carbon, B. A., Bartle, G. A., Murray, A. M., and MacPherson, D. K. (1980). The distribution of root length, and the limits to flow of soil water to roots in a dry sclerophyll forest. *Forest Sci.* **26**, 656–664.
Charles-Edwards, D. A. (1981). "The Mathematics of Photosynthesis and Productivity" Academic Press, London.
Charles-Edwards, D. A., and Thorpe, M. R. (1976). Interception of diffuse and direct-beam radiation by a hedgerow apple orchard. *Ann. Bot.* **40**, 603–613.
Charley, J. (1981). Soils as a nutrient reservoir and buffer system. *In* "Australian Forest Nutrition Workshop: Productivity in Perpetuity" pp. 13–27. CSIRO, Melbourne.
Christersson, L. and Sandstedt, R. (1978). Short-term temperature variation in needles on *Pinus sylvestris* L. *Can. J. For. Res.* **8**, 480–482.
Collatz, J., Ferrar, P. J., and Slatyer, R. O. (1976). Effects of water stress and differential hardening treatments on photosynthetic characteristics of a xeromorphic shrub *Eucalyptus socialis* F. Muell. *Oecologia (Berl)*. **23**, 95–105.
Cotterill, P. P., and Nambiar, E. K. S. (1981). Seedling physiology of three radiata pine families with parents of contrasting growth. *Aust. For. Res.* **11**, 13–22.
Cowan, I. R. (1968). Mass, heat and momentum exchange between stands of plants and their atmospheric environment. *Quart. J. R. Met. Soc.* **94**, 523–544.
Cremer, K. W. (1976). Daily patterns of shoot elongation in *Pinus radiata* and *Eucalyptus regnans*. *New Phytol.* **76**, 459–468.
Cromer, R. N. and Williams, E. R. (1982). Biomass and nutrient accumulation in a planted *E. globulus* (habill) fertilizer trial. *Aust. J. Bot.* **30**, 265–278.
Cromer, R. N., Tompkins, D., and Barr, N. J. (1983). Irrigation of *Pinus radiata* with waste water: tree growth in response to treatment. *Aust. For. Res.* **13**, 57–65.
Daniel, T. W., Helms, J. A., and Baker, F. S. (1979). Principles of silviculture. 2nd ed McGraw-Hill, New York.
Davey, C. B. and Wollum, A. G. (1979). Nitrogen inputs to forest ecosystems through biological fixation. *In* "Impact of Intensive Harvesting on Forest Nutrient Cycling" USDA Forest Service, State University of New York, pp. 62–74.
Davidson, R. L. (1969). Effects of root-leaf temperature differentials on root/shoot ratios on some pasture grasses and clover. *Ann. Bot.* **33**, 561–569.
Deans, J. D. (1979). Fluctuations of the soil environment and fine root growth in a young Sitka spruce plantation. *Plant and Soil* **52**, 195–208.
Deans, J. D. (1981). Dynamics of coarse root production in a young plantation of *Picea sitchensis*. *Forestry* **54**, 139–153.
Denmead, O. T. (1984). Plant physiological methods for studying evapotranspiration: problems of telling the forest from the trees. *Agricultural Water Management* **8**, 187–189.
de Vries, D. A. (1958). Simultaneous transfer of heat and moisture in porous media. *Trans. Am. Geophys. Un.* **39**, 909–916.
de Wit, C. T. (1970). Dynamic concepts in biology. *In* "Prediction and Measure-

ment of Photosynthetic Productivity" (I. Setlik, ed), pp. 17–23. PUDOC, Wageningen.

Dilley, A. C. (1968). On the computer calculation of vapour pressure and specific humidity gradients from psychrometric data. *J. Appl. Meterol.* **7**, 717–719.

Doley, D. (1981). Tropical and sub-tropical forests and woodlands. In "Water Deficits and Plant Growth" Vol. VI (T. T. Kozlowski, ed.), pp. 209–307. Academic Press, New York.

Dosskey, M. G., and Ballard, T. M. (1980). Resistance to water uptake by Douglas-fir seedlings in soils of different texture. *Can. J. For. Res.* **10**, 530–534.

Drew, T. J., and Flewelling, J. W. (1977). Some recent Japanese theories of yield-density relationships and their application to Monterey pine plantations. *Forest Sci.* **4**, 517–534.

Ehleringer, J. R. and Björkman, O. (1977). Quantum yields for CO_2 uptake in C_3 and C_4 plants: dependence on temperature, CO_2, and O_2 concentration. *Plant Physiol.* **59**, 86–90.

Farquhar, G. D., and Sharkey, T. D. (1982). Stomatal conductance and photosynthesis. *Ann. Rev. Plant Physiol.* **33**, 317–345.

Farquhar, G. D., and von Caemmerer, S. (1982). Modelling of photosynthetic response to environmental conditions. In "Encyclopaedia of Plant Physiology Vol. 12 B. Physiological Plant Ecology II: Water Relations and Carbon Assimilation" (O. L. Lange, P. S. Nobel, C. B. Osmond, H. Ziegler eds.), pp. 549–587. Springer Verlag, Berlin.

Farquhar, G. D., von Caemmerer, S., and Berry, J. A. (1980). A biochemical model of photosynthetic CO_2 assimilation in leaves of C_3 species. *Planta* **149**, 78–90.

Federer, C. A. (1968). Spatial variation of net radiation, albedo and surface temperature of forests. *J. Appl. Meteorol.* **7**, 789–795.

Fife, D. N., and Nambiar, E. K. S. (1982). Accumulation and retranslocation of mineral nutrients in developing needles in relation to seasonal growth in young radiata pine trees. *Ann. Bot.* **50**, 817–829.

Fife, D. N., and Nambiar, E. K. S. (1984). Movement of nutrients in radiata pine needles in relation to the growth of shoots. *Ann. Bot.* **54**, 303–314.

Ford, E. D., and Deans, J. D. (1977). Growth of a Sitka spruce plantation: spatial distribution and seasonal fluctuations of lengths, weights and carbohydrate concentrations of fine roots. *Plant and Soil* **47**, 463–485.

Forrest, W. G. (1969). Variations in the accumulation, distribution and movement of mineral nutrients in radiata pine plantations. Ph.D. thesis, Australian National University.

Fowkes, N. D., and Landsberg, J. J. (1981). Optimal root systems in terms of water uptake and movement. In "Mathematics and Plant Physiology" (D. A. Rose, and D. A. Charles-Edwards, eds.), pp. 109–125. Academic Press, London.

Fujimori, T., Kawanabe, S., Saito, H., Grier, C. C., and Shidei, T. (1976). Biomass and primary production in forests of three major vegetation zones of the northwestern United States. *J. Jap. For. Soc.* **58**, 360–373.

Gallagher, J. N., and Biscoe, P. V. (1978). Radiation absorption, growth and yield of cereals. *J. Agric. Sci.* **91**, 47–60.

Gardner, W. R. (1958). Same steady state solutions of the unsaturated moisture flow equation with application to evaporation from a water table. *Soil Sci.* **85**, 228–232.

Gardner, W. R., Hillel, D., and Benjamini, Y. (1970). Post-irrigation movement of soil water: I. Redistribution. *Water Resources Res.* **6**, 851–886.

Garratt, J. R. (1977). Aerodynamic roughness and mean monthly surface stress over Australia. CSIRO, Australia. Division of Atmospheric Physics, Tech. paper No. 29.

Gash, J. H. C. (1979). An analytical model of rainfall interception by forests. *Quart. J.R. Met. Soc.* **105**, 43–55.

Gash, J. H. C., and Morton, A. J. (1978). An application of the Rutter model to the estimation of the interception loss from Thetford forest. *J. Hydrol.* **38**, 49–58.

Gerwitz, A., and Page, E. R. (1974). An empirical mathematical model to describe plant root systems. *J. Appl. Ecol.* **11**, 773–782.

Gholz, H. L., Grier, C. C., Campbell, A. G., and Brown, A. T. (1979). Equations for estimating biomass and leaf area of plants in the Pacific-northwest. Research Paper No. 41, Oregon State University School of Forestry, Corvallis, Oregon.

Gordon, J. C., and Larson, P. R. (1968). Seasonal course of photosynthesis, respiration and distribution of ^{14}C in young *Pinus resinosa* trees as related to wood formation. *Plant Physiol.* **43**, 1617–1624.

Grace, J. (1977). "Plant response to wind". Academic Press, London

Grace, J. (1981). Some effects of wind on plants. *In* "Plants and their Atmospheric Environment" (J. Grace, E. D. Ford, and P. G. Jarvis, eds), pp. 31–56. Blackwell, Oxford.

Grace, J. (1983). Plant–atmosphere relationships. Chapman and Hall, London and New York.

Graham, R. D. (1984). Breeding for nutritional characteristics in cereals. *In* "Advances in Plant Nutrition" Academic Press, New York.

Greacen, E. L. and Hignett, C. J. (1984). Water balance under wheat modelled with limited soil data. *Agricultural Water Management.* **8**, 291–304.

Grier, C. C., and Running, S. W. (1977). Leaf area of mature northwestern coniferous forests: relation to site water balance. *Ecology* **58**, 893–899.

Grier, C. C., and Waring, R. H. (1974). Conifer foliage mass related to sapwood area. *Forest Sci.* **20**, 205–206.

Grier, C. C., Vogt, K. A., Keyes, M. R., and Edmonds, R. L. (1981). Biomass distribution and above- and below-ground production in young and mature *Abies amabilis* zone ecosystems of the Washington Cascades. *Can. J. For. Res.* **11**, 155–167.

Hälldin, S. (ed.) (1979). "Comparison of forest water and energy exchange models" Int. Soc. for Ecological Modelling, Copenhagen, Denmark.

Havranek, W. M. (1981). Stammatmung, dickenwachstum und photosynthese einer zirbe (*Pinus cembra*, L) an der Walgrenze. *Mitt. Forstl. Bundesvers. Wien* **142**, 443–467.

Helkvist, J., Richards, G. P., and Jarvis, P. G. (1974). Vertical gradients of water potential and tissue water relations in Sitka spruce trees measured with the pressure chamber. *J. Appl. Ecol.* **11**, 637–668.

Herkelrath, W. N., Miller, E. E., and Gardner, W. R. (1977). Water uptake by plants: II The root contact model. *Soil Sci. Soc. Am.* **41**, 1039–1043.

Herwitz, S. R. (1982). The redistribution of rainfall by tropical rainforest canopy tree species. *In* "First National Symposium on Forest Hydrology" (E. M. O'Loughlin and L. J. Bren, eds.), pp. 26–29. Institution of Engineers, Australia.

Hillel, D. (1980a). "Fundamentals of soil physics." Academic Press, New York.
Hillel, D. (1980b). "Applications of soil physics." Academic Press, New York.
Hinckley, T. M., and Bruckerhoff, D. N. (1975). The effects of drought on water relations and stem shrinkage of *Quercus alba*. *Can. J. Bot.* **53**, 62–72.
Hinckley, T. M., Lassoie, J. P., and Running, S. W. (1978). Temporal and spatial variations in the water status of forest trees. Forest Science, Monograph 20. Society of American Foresters.
Holden, J. M., Thomas, G. W. and Jackson, R. M. (1983). Effect of mycorrhizal inoculae on the growth of Sitka spruce seedlings in different soils. *Plant and Soil* **71**, 313–317.
Hsiao, T. (1973). Plant responses to water stress. *Ann. Rev. Plant Physiol.* **24**, 519–570.
Ingestad, T. (1982). Relative addition rate and external concentration; driving variables used in plant nutrition research. *Plant Cell and Environ.* **5**, 443–453.
Jackson, D. S., and Chittenden, J. (1981). Estimation of dry matter in *Pinus radiata* root systems: I. Individual trees. *N.Z. J. For. Sci.* **11**, 164–182.
Jackson, J. E., and Palmer, J. W. (1979). A simple model of light transmission and interception by discontinuous canopies. *Ann. Bot.* **44**, 381–383.
Jarvis, P. G. (1971). The estimation of resistances to carbon dioxide transfer. *In* "Plant Photosynthetic Production: Manual of Methods". (F. Sestak, J. Catsky, and P. G. Jarvis, eds.), pp. 566–631. Dr J. Junk Publishers, The Hague.
Jarvis, P. G. (1975). Water transfer in plants. *In* "Heat and Mass Transfer in the Plant Environment, Part 1". (H. D. A. de Vries, and N. G. Afgan, eds.), pp. 369–394. Scripta Book Co., Washington, D.C.
Jarvis, P. G. (1976). The interpretation of the variations in leaf water potential and stomatal conductance found in canopies in the field. *Phil. Trans. R. Soc. Ser. B* **273**, 593–610.
Jarvis, P. G. (1981). Production efficiency of coniferous forest in the UK. *In* "Physiological Processes Limiting Plant Productivity" (C. B. Johnson, ed.), pp. 81–107. Butterworth, London.
Jarvis, P. G., and Mansfield, T. A. (eds.), (1981). "Stomatal physiology." Cambridge University Press, Cambridge.
Jarvis, P. G. and Leverenz, J. W. (1983). Productivity of temperate, deciduous and evergreen forests. *In* "Encyclopaedia of Plant Physiology, Vol. 12D. Physiological Plant Ecology IV. Ecosystem Processes: Mineral cycling, Productivity and Man's influence" (O. L. Lange, P. S. Nobel, C. B. Osmond, and H. Zeigler, eds.), pp. 233–280. Springer Verlag, Berlin.
Jarvis, P. G., James, G. B., and Landsberg, J. J. (1976). Coniferous forest. *In* "Vegetation and the Atmosphere" Vol. 2 (J. L. Monteith, ed.), pp. 171–264. Academic Press, New York.
Jarvis, P. G., Edwards, W. R. N., and Talbot, H. (1981). Models of plant and crop water use. *In* "Mathematics and Plant Physiology" (D. A. Rose and D. A. Charles–Edwards, eds), pp. 151–194. Academic Press, London and Orlando.
Jones, H. G. (1978). Modelling diurnal trends of leaf water potential in transpiring wheat. *J. Appl. Ecol.* **15**, 613–626.
Jones, H. G. (1983a). Estimation of an effective soil water potential at the root surface of transpiring plants. *Plant, Cell and Environ.* **6**, 671–674.
Jones, H. G. (1983b). Plants and microclimate. Cambridge University Press, Cambridge.
Kalma, J. P., and Fuchs, M. (1976). Citrus orchards. *In* "Vegetation and the

Atmosphere" Vol. 2 (J. L. Monteith, ed.), pp. 309–329. Academic Press, London.
Keyes, M. R., and Grier, C. C. (1981). Above- and below-ground net production in 40-year-old Douglas-fir stands on low and high productivity sites. *Can. J. For. Res.* **11,** 599–605.
Kinerson, R. J., Higginbotham, K. O., and Chapman, R. C. (1974). Dynamics of foliage distribution within a forest canopy. *J. Appl. Ecol.* **11,** 347–353.
Korner, Ch., Scheel, J. A., and Bauer, H. (1979). Maximum leaf diffusive conductance in vascular plants. *Photosynthetica* **13,** 45–82.
Kozlowski, T. T. (ed.) (1981). "Water deficits and plant growth." Vol. VI. Academic Press, New York.
Kozlowski, T. T. (1982). Water supply and tree growth I. Water deficits. *For. Abs.* **43,** 57–161.
Landsberg, J. J. (1974). Apple fruit bud development and growth: analysis and an empirical model. *Ann. Bot.* **38,** 1013–1023.
Landsberg, J. J. (1977). Some useful equations for biological studies. *Expl. Agric.* **13,** 272–286.
Landsberg, J. J. (1981a). The use of models in interpreting plant response to weather. *In* "Plants and their Atmospheric Environments" (J. Grace, E. D. Ford, P. G. Jarvis, eds.), pp. 369–389. Blackwell, Oxford.
Landsberg, J. J. (1981b). The number and quality of the driving variables needed to model tree growth. *Studia Forestalia Suecica* **160,** 43–50.
Landsberg, J. J. (1984). Physical aspects of the water regime of wet tropical vegetation. *In* "Physiological Ecology of Plants of the Wet Tropics" (E. Medina, H. A. Mooney and C. Vazquez-Yanes, eds.), pp. 13–25. Dr W. Junk Publishers, The Hague.
Landsberg, J. J. and Butler, D. R. (1980). Stomatal response to humidity: implications for transpiration. *Plant Cell and Environ.* **3,** 29–33.
Landsberg, J. J. and Fowkes, N. D. (1978). Water movement through plant roots. *Ann. Bot.* **42,** 493–508.
Landsberg, J. J. and James, G. B. (1971). Wind profiles in plant canopies: studies on an analytical model. *J. Appl. Ecol.* **8,** 729–741.
Landsberg, J. J. and Jarvis, P. G. (1973). A numerical investigation of the momentum balance of a spruce forest. *J. Appl. Ecol.* **10,** 645–655.
Landsberg, J. J., and McMurtrie, R. (1984). Water use by isolated trees. *Agricultural Water Management* **8,** 223–242.
Landsberg, J. J., and Powell, D. B. B. (1973). Surface exchange characteristics of leaves subject to mutual interference. *Agric. Meteorol.* **12,** 169–189.
Landsberg, J. J., and Thom, A. S. (1971). Aerodynamic properties of a plant of complex structure. *Quart. J. R. Met. Soc.* **97,** 565–570.
Landsberg, J. J., Beadle, C. L., Biscoe, P. V., Butler, D. R., Davidson, B., Incoll, L. D., James, G. B., Jarvis, P. G., Martin, P. J., Neilson, R. E., Powell, D. B. B., Slack, Elisabeth M., Thorpe, M. R., Turner, N. C., Warrit, B., and Watts, W. R. (1975). Diurnal energy, water and CO_2 exchanges in an apple (*Malus pumila*) orchard. *J. Appl. Ecol.* **12,** 659–684.
Landsberg, J. J., Blanchard, T. W., and Warrit, B. (1976). Studies on the movement of water through apple trees. *J. Exp. Bot.* **27,** 579–596.
Lang, A. and Thorpe, M. R. (1983). Analysing partitioning in plants. *Plant, Cell and Environ.* **6,** 267–274.
Langenheim, J. H., Osmond, C. B., Brooks, A. and Ferrar, P. J. (1984). Photosynthetic responses to light in seedings of selected Amazonian and Australian rainforest tree species. *Oecologia* (Berl.) **63,** 215–224.

Larsson, S. and Bengston, C. (1980). Effects of water stress on growth in Scots pine. Technical report 24, Swedish Coniferous Forest project.

Larson, P. R. and Gordon, J. C. (1969). Leaf development, photosynthesis, and C^{14} distribution in *Populus deltoides* seedlings. *Am. J. Bot.* **56**, 1058–1066.

Leonard, R. E. and Federer, C. A. (1973). Estimated and measured roughness parameters for a pine forest. *J. Appl. Meteorol.* **12**, 302–307.

Leuning, R. and Attiwill, P. M. (1978). Mass, heat and momentum exchange between a moist eucalyptus forest and the atmosphere. *Agric. Meteorol.* **19**, 215–214.

Linder, S. (1985). Potential and actual production in Australian forest stands. *In* "Research for Forest Management". (J. J. Landsberg and W. Parsons, eds.), pp. 11–35. CSIRO, Melbourne.

Linder, S. and Axelsson, B. (1982). Changes in carbon uptake and allocation patterns as a result of irrigation and fertilization in a young *Pinus sylvestris* stand. *In* "Carbon Uptake and Allocation in Subalpine Ecosystems as a Key to Management". (R. H. Waring, ed.), pp. 38–44. Forest Research Laboratory, Oregon State University.

Linder, S. and Rook, D. A. (1984). Effects of mineral nutrition on carbon dioxide exchange and partitioning of carbon in trees. *In* "Nutrition of Plantation Forests". (G. D. Bowen and E. K. S. Nambiar, eds), 211–236. Academic Press, London.

Linder, S. and Troeng, E. (1981). The seasonal variation in stem and coarse root respiration of a 20-year-old Scots pine (*Pinus sylvestris* L.). *Mitt. Forstl. Bundes Vers. Wien*.

List, R. J. (1968). Smithsonian Meteorological Tables. Smithsonian Institute Press, Washington.

Ludlow, M. M. and Jarvis, P. G. (1971). Photosynthesis in Sitka spruce (*Picea sitchensis* (Bong.) Carr.) I. General characteristics. *J. Appl. Ecol.* **8**, 925–953.

Luvall, J. C. and Murphy, C. F. (1982). Evaluation of the tritiated water method for measurement of transpiration in young *Pinus taeda* L. *Forest Sci.* **28**, 5–16.

Madgwick, H. A. I., Jackson, D. S. and Knight, P. J. (1977). Above-ground dry matter, energy and nutrient contents of trees in an age series of *Pinus radiata* plantations. *N.Z. J. For. Sci.* **7**, 445–468.

Mayr, E. (1983). The growth of biological thought. Harvard University Press, Cambridge, MA.

McClaugherty, C. A., Aber, J. D. and Melillo, J. M. (1982). The role of fine roots in the organic matter and nitrogen budgets of two forested ecosystems. *Ecology* **63**, 1481–1490.

McCree, K. J. (1970). An equation for the rate of respiration of white clover plants grown under controlled conditions. *In* "Prediction of Measurement of Photosynthetic Productivity" (I. Setlik, ed.), pp. 332–339. PUDOC, Wageningen.

McCree, K. J. (1974). Equations for the rate of dark respiration of white clover and grain sorghum, as a function of dry weight, photosynthetic rate and temperature. *Crop Sci.* **14**, 509–514.

McCree, K. J. (1981). Photosynthetically active radiation. *In* "Encyclopaedia of Plant Physiology, Vol. 12A. Physiological plant ecology I. Responses to the Physical Environment" (O. L. Lange, P. S. Nobel, C. B. Osmond and H. Ziegler, eds), pp. 41–55. Springer Verlag, Berlin.

McMurtrie, R. and Wolf, L. (1983). A model of competition between trees and grass for radiation, water and nutrients. *Ann. Bot.* **52**, 449–458.

McNaughton, K. G. and Black, T. A. (1973). A study of evapotranspiration from a Douglas-fir forest using the energy balance approach. *Water Resources Res.* **9**, 1579–1590.

McNaughton, K. G. and Jarvis, P. G. (1983). Predicting effects of vegetation changes on transpiration and evaporation. In "Water Deficits and Plant Growth" Vol. VII (T. T. Kozlowski, ed.), pp. 1–47. Academic Press, New York.

Miller, H. G. (1981). Nutrient cycles in forest plantations, their change with age and the consequence for fertilizer practice. In "Australian Forest Nutrition Workshop: Productivity in Perpetuity" pp. 181–199. CSIRO, Melbourne.

Miller, H. G. (1984). Dynamics of nutrient cycling in plantation ecosystems. In "Nutrition of Plantation Forests". (G. D. Bowen and E. K. S. Nambiar, eds), pp. 53–78. Academic Press, London.

Miller, H. G., Cooper, J. M. and Miller, J. D. (1976a). Effect of nitrogen supply on nutrients in litter fall and crown leaching in a stand of Corsican pine. *J. Appl. Ecol.* **13**, 233–248.

Miller, H. G., Miller, J. D. and Pauline, O. J. L. (1976b). Effect of nitrogen supply on nutrient uptake in Corsican pine. *J. Appl. Ecol.* **13**, 955–966.

Milthorpe, F. L., and Moorby, J. (1974, 1979). An Introduction to Crop Physiology. 1st and 2nd eds. Cambridge University Press, Cambridge.

Monteith, J. L. (1973). Principles of environmental physics. 1st edition. Edward Arnold, London.

Monteith, J. L. (1981). Climatic variation and growth of crops. *Quart. J. R. Met. Soc.* **107**, 749–774.

Moore, C. J. (1976). A comparative study of radiation balance above forest and grassland. *Quart. J. R. Met. Soc.* **102**, 889–899.

Nambiar, E. K. S., Cotterill, P. P. and Bowen, G. D. (1982). Genetic differences in the root regeneration of radiata pine. *J. Exp. Bot.* **33**, 170–177.

Norman, J. M. (1979). Modelling the complete crop canopy. In "Modification of the Aerial Environment of Crops". (B. J. Barfield and J. F. Gerber, eds), pp. 249–277. Am. Soc. Agric. Engineers.

Norman, J. M. (1982). Simulation of microclimates. In "Biometeorology in Integrated Pest Management" (J. L. Hatfield and I. J. Thomason, eds), pp. 65–99. Academic Press, New York.

Nye, P. H. and Tinker, P. B. (1978). "Solute movement in the soil–root system". Blackwell, Oxford.

O'Connell, A. M., Grove, T. S., and Lamb, D. (1981). The influence of fire on the nutrition of Australian forests. In "Australian Forest Nutrition Workshop: Productivity in Perpetuity", pp. 277–289. CSIRO, Melbourne.

Oke, T. R. (1978). "Boundary Layer Climates". Methuen, London.

Oquist, G., Brunes, L. and Hallgren, J. E. (1982). Photosynthetic efficiency of *Betula pendula* acclimated to different quantum flux densities. *Plant, Cell and Environ.* **5**, 9–15.

Osmond, C. B., Björkman, O. and Anderson, D. J. (1980). "Physiological processes in plant ecology." Springer Verlag, Berlin.

Ovington, J. D. (1957). Dry matter production in *Pinus sylvestris*. *Ann. Bot.* **21**, 257–314.

Parker, W. C., Pallardy, S. G., Hinckley, T. M. and Teskey, R. O. (1982). Seasonal changes in tissue water relations of three woody species of the *Quercus-carya* forest type. *Ecology* **63**, 1259–1267.

Pastor, J. and Bockheim, J. G. (1984). Distribution and cycling of nutrients in an aspen-mixed-hardwood-spodosol ecosystem in northern Wisconsin. *Ecology* **65**, 339–353.

Pastor, J., Aber, J. D., McClaugherty, C. A. and Melillo, J. M. (1984). Aboveground production and N and P cycling along a nitrogen mineralization gradient on Blackhawk Island, Wisconsin. *Ecology* **65**, 256–268.

Passioura, J. B. (1982). Water in the soil-plant-atmosphere continuum. *In* "Encyclopaedia of Plant Physiology, Vol. 12 B. Physiological Plant Ecology II: Water Relations and Carbon Assimilation" (O. L. Lange, P. S. Nobel, E. B. Osmond, H. Ziegler, eds), pp. 5–33. Springer Verlag, Berlin.

Pearcy, R. W., Osteryoung, K. and Randall, D. (1982). Carbon dioxide exchange characteristics of C_4 Hawaiian *Euphorbia* species native to diverse habitats. *Oecologia* (Berl.) **55**, 333–341.

Pearson, J. A., Fahey, T. J. and Knight, D. H. (1984). Biomass and leaf area in contrasting lodgepole pine forests. *Can. J. For. Res.* **14**, 259–264.

Persson, H. (1978). Root dynamics in a young Scots pine stand in central Sweden. *Oikos* **30**, 508–519.

Persson, H. (1980a). Spatial distribution of fine-root growth, mortality and decomposition in a young Scots pine stand in Central Sweden. *Oikos* **34**, 77–87.

Persson, H. (1980b). Fine-root dynamics in a Scots Pine stand with and without near-optimum nutrient and water regimes. *Acta Phytogeogr. Suec.* **68**, 101–110.

Pinker, R. T. (1982). The diurnal asymmetry in the albedo of tropical forest vegetation. *Forest Sci.* **28**, 297–304.

Pook, E. W. (1984a). Canopy dynamics of *Eucalyptus maculata* Hook. I. Distribution and dynamics of leaf populations. *Aust. J. Bot* **32**, 387–403.

Pook, E. W. (1984b). Canopy dynamics of *Eucalyptus maculata* Hook. II. Canopy leaf area balance. *Aust. J. Bot.* **32**, 405–413.

Powell, D. B. B. and Thorpe, M. R. (1977). Dynamic aspects of plant-water relations. *In* "Environmental effects on Crop Physiology" (J. J. Landsberg, C. V. Cutting, eds), pp. 259–285. Academic Press, London.

Pritchett, W. L. (1979). Soil as a reservoir: soil nutrient supplies and mobilization rate. *In* "Impact of Intense Harvesting on Forest Nutrient Cycling". USDA Forest Service, State University of New York, pp. 49–61.

Raison, R. J. (1979). Modification of the soil environment by vegetation fires, with particular reference to nitrogen transformations: a review. *Plant and Soil* **51**, 73–108.

Raison, R. J. (1984). Potential adverse effects of forest operations on the fertility of soils supporting fast-growing plantations. *In* "Site and Productivity of Fast-growing Plantations", pp. 457–472. Proc. IUFRU Symp. Pretoria/Pietermaritzburg, South Africa.

Raison, R. J., Khanna, P. K. and Woods, P. V. (1985). Mechanisms of element transfer to the atmosphere during vegetation fires. *Can. J. For. Res.* **15**, 132–140.

Rauner, H. L. (1976). Deciduous forests. *In* "Vegetation and the Atmosphere" Vol. 2 (J. L. Monteith, ed.), pp. 241–264. Academic Press, London.

Reich, P. G. and Borchert, R. (1982). Phenology and ecophysiology of the tropical tree, *Tabeonia neochrysantha* (Bignomiaceae). *Ecology* **63**, 294–299.

Richter, H. (1973). Frictional potential losses and total water potential in plants: a re-evaluation. *J. Exp. Bot.* **27**, 473–479.

Rook, D. A. and Corson, M. J. (1978). Temperature and irradiance and the total daily photosynthetic production of the crown of a *Pinus radiata* tree. *Oecologia* (Berl.) **36**, 371–382.

References

Ross, J. (1976). Radiative transfer in plant communities. *In* "Vegetation and the Atmosphere" Vol. I (J. L. Monteith, ed.), pp. 13–55. Academic Press, London.
Rundell, P. W. (1981). Fire as an ecological factor. *In* "Encyclopaedia of Plant Physiology, Vol. 12 A. Physiological plant ecology I: Responses to the physical environment". (O. L. Lange, P. S. Nobel, C. B. Osmond, H. Ziegler, eds), pp. 501–538. Springer Verlag, Berlin.
Running, S. W. (1980). Field estimates of root and xylem resistance in *Pinus contorta* using root excision. *J. Exp. Bot.* **31**, 555–569.
Rutter, A. J., Kershaw, K. A., Robins, P. L. and Morton, A. J. (1971). A predictive model of rainfall interception in forests. I. Derivation of the model from observations in a plantation of Corsican pine. *Agric. Meteorol.* **9**, 367–383.
Rutter, A. J., Morton, A. J. and Robins, P. L. (1975). A predictive model of rainfall interception in forests. II. Generalization of the model and comparison with observations in some coniferous and hardwood stands. *J. Appl. Ecol.* **12**, 367–380.
Satoo, T., and Madgwick, H. A. I. (1982). "Forest biomass". Martinus Nijhoff, The Hague.
Schulze, E. D. and Hall, A. E. (1982). Stomatal responses, water loss and CO_2 assimilation rates of plants in contrasting environments. *In* "Encyclopaedia of Plant Physiology, Vol. 12 B. Physiological Plant Ecology II: Water Relations and Carbon Assimilation" (O. L. Lange, P. S. Nobel, C. B. Osmond, H. Ziegler, eds), pp. 181–230. Springer Verlag, Berlin.
Sellers, W. D. (1965). Physical climatology. University of Chicago Press, Chicago, II.
Siau, J. F. (1971). "Flow in Wood." Syracuse Wood Sci. Ser. I. Syracuse University Press, Syracuse, NY.
Sinclair, R. (1980). Water potential and stomatal conductance of three *Eucalyptus* species in the Mount Lofty ranges, South Australia: responses to summer drought. *Aust. J. Bot.* **28**, 499–510.
Skaar, C. (1972). "Water in Wood". Syracuse Wood Sci. Ser. 4. Syracuse University Press, Syracuse, NY.
Slatyer, R. O. (1967). "Plant water relationships." Academic Press, London.
Stanhill, G. (1970). Some results of helicopter measurements of the albedo of different land surfaces. *Solar Energy* **13**, 59–66.
Stewart, J. B. and Thom, A. S. (1973). Energy budgets in a pine forest. *Quart. J. R. Met. Soc.* **99**, 154–170.
Stewart, H. T. L., Flinn, D. W. and James, J. M. (1981). Biomass and nutrient distribution in radiata pine. *In* "Australian Forest Nutrition Workshop: Productivity in Perpetuity", pp. 173–185. CSIRO, Melbourne.
Stigter, C. J. (1980). Solar radiation as statistically related to sunshine duration: a comment using low-latitude data. *Agric. Meteorol.* **21**, 173–178.
Switzer, G. L. and Nelson, L. E. (1972). Nutrient accumulation and cycling in Loblolly pine (*Pinus taeda* L.) plantation ecosystems: the first twenty years. *Soil Sci. Soc. Am. Proc.* **36**, 143–147.
Tan, C. S., Black, T. A. and Nayamah, J. U. (1978). A simple diffusion model of transpiration applied to a thinned Douglas-fir stand. *Ecology* **59**, 1221–1229.
Tajchman, S. J. (1972). The radiation and energy balances of coniferous and deciduous forests. *J. Appl. Ecol.* **9**, 359–375.
Tajchman, S. J. (1981). Comments on measuring turbulent exchange within and above forest canopy. *Bull. Am. Meteorol. Soc.* **62**, 1550–1559.

Tajchman, S. J. (1984). Distribution of the radiation index of dryness and forest site quality in a mountainous watershed. *Can. J. For. Res.* **14**, 717–721.

Thom, A. S. (1971). Momentum absorption by vegetation. *Quart. J. R. Met. Soc.* **97**, 414–428.

Thorpe, M. R., Warrit, B. and Landsberg, J. J. (1980). Responses of apple leaf stomata: a model for single leaves and a whole tree. *Plant, Cell and Environ.* **3**, 23–27.

Troeng, E. and Linder, B. S. (1982). Gas exchange in a 20-year-old stand of Scots pine. II. Variation in net photosynthesis within and between trees. *Physiol. Plant.* **54**, 15–23.

Turner, J. (1981). Nutrient supply in relation to immobilization in biomass and nutrient removal in harvesting. *In* "Australian Forest Nutrition Workshop: Productivity in Perpetuity" pp. 263–275. CSIRO, Melbourne.

Turner, N. C. and Jones, M. M. (1980). Turgor maintenance by osmotic adjustments; a review and evaluation. *In* "Adaptation of Plants to Water and High Temperature Stress" (N. C. Turner, P. J. Kramer, eds), pp. 87–104. Wiley, New York.

Tyree, M. T. and Jarvis, P. G. (1982). Water in tissues and cells. *In* "Encyclopaedia of Plant Physiology, Vol. 12B. Physiological Plant Ecology II. Water Relations and Carbon Assimilation" (O. L. Lange, P. S. Nobel, C. B. Osmond, H. Ziegler, eds), pp. 35–77. Springer Verlag, Berlin.

Walker, J., Raison, R. J., and Khanna, P. K. (1986). The impact of fire on Australian soils. *In* "The Impact of Man on Australian Soils". (J. S. Russell and R. F. Isbell, eds), in press. University of Queensland Press, Queensland, Australia.

Wallace, J. S. and Biscoe, P. V. (1983). Water relations of winter wheat: 4. Hydraulic resistance and capacitance in the soil-plant system. *J. Agric. Sci.* **100**, 591–589.

Wallace, J. S., Clark, J. A. and McGowan, M. (1983). Water relations of winter wheat: 3. Components of leaf water potential and the soil-plant water potential gradient. *J. Agric. Sci.* **100**, 581–589.

Waring, R. H. and Edmonds, R. L. (eds). (1974). Integrated Research in the Coniferous Forest Biome Ecosystem Analysis Studies, US Biological Program.

Waring, R. H. and Running, S. W. (1978). Sapwood water storage: its contribution to transpiration and effect upon water conductance through the stems of old-growth Douglas-fir. *Plant, Cell and Environ.* **1**, 131–140.

Waring, R. H., Newman, K. and Bell, J. (1981). Efficiency of tree crowns and stemwood production at different canopy leaf densities. *Forestry*, **54**, 129–137.

Waring, R. H., Whitehead, D. and Jarvis, P. G. (1979). The contribution of stored water to transpiration in Scots pine. *Plant, Cell and Environ.* **2**, 309–317.

Waring, R. H., Whitehead, D. and Jarvis, P. G. (1980). Comparison of an isotopic method and the Penman-Monteith equation for estimating transpiration from Scots pine. *Can. J. For. Res.* **10**, 555–558.

Watson, R. L. and Landsberg, J. J. (1979). The photosynthetic characteristics of apple leaves (c.v. Golden Delicious) during their early growth. *In* "Photosynthesis and Plant Development" (R. Marcelle, H. Clisters and M. van Poucke, eds), pp. 39–48. Dr W. Junk, Publishers, The Hague.

Watson, R. L., Landsberg, J. J. and Thorpe, M. R. (1978). Photosynthetic characteristics of the leaves of "Golden Delicious" apple trees. *Plant, Cell and Environ.* **1**, 51–58.

Watts, W. R., Neilson, R. E. and Jarvis, P. G. (1976). Photosynthesis in Sitka spruce

(*Picea sitchensis* (Bong). Carr) VII. Measurements of stomatal conductance and CO_2 uptake in a forest canopy. *J. Appl. Ecol.* **13**, 623–638.

White, J. (1981). The allometric interpretation of the self-thinning rule. *J. Theor. Biol.* **89**, 475–500.

Whitehead, D. (1978). The estimation of foliage area from basal area in Scots pine. *Forestry* **51**, 137–149.

Whitehead, D. (1985) A review of processes in the water relations of forests. *In* "Research for Forest Management" (J. J. Landsberg and W. Parsons, eds), pp. 94–124. CSIRO, Melbourne.

Whitehead, D. and Jarvis, P. G. (1981). Coniferous forests and plantations. *In* "Water Deficits and Plant Growth". Vol. 6. (T. T. Kozlowski, ed.), pp. 49–152. Academic Press, New York.

Whitehead, D., Okali, D. U. U. and Fasehun, F. E. (1981). Stomatal response to environmental variables in two tropical forest species during the dry season in Nigeria. *J. Appl. Ecol.* **18**, 571–587.

Whitehead, D., Jarvis, P. G. and Waring, R. H. (1984). Stomatal conductance, transpiration and resistance to water uptake in a *Pinus sylvestris* spacing experiment. *Can. J. For. Res.* **14**, 692–700.

Whittaker, R. H. and Woodwell, G. M. (1967). Surface area relations of woody plants and forest communities. *Am. J. Bot.* **54**, 931–939.

Wong, S. C., Cowan, I. R., and Farquhar, G. D. (1978). Leaf conductance in relation to assimilation in *Eucalyptus pauciflora*. Sieb ex Spreng. *Plant Physiol.* **61**, 670–674.

Wong, S. C., Cowan, I. R., and Farquhar, G. D. (1979). Stomatal conductance correlates with photosynthetic capacity. *Nature* **282**, 424–426.

Wood, T., Bormann, F. H. and Voigt, G. K. (1984). Phosphorus cycling in a northern hardwood forest: biological and chemical control. *Science* **223**, 391–393.

Index

Abies amabilis, 99, 107
Absorbed energy, 44
  radiation, 174
Absolute humidity, 26
Acer macrophyllum, 99
Acclimation, 158
Acceleration due to gravity, 146
Adaptation, 158
Adiabatic lapse rate, 55
Aerodynamic characteristics, 59
  exchange processes, 31
  properties of surfaces, 57
Agathis microstachya, 84
Agathis robusta, 84
Air density, 45
  humidity, 25, 32
  specific heat, 45
  temperature, 22, 23, 24
Air flow in plant communities, 60
Albedo, 43, 44
  bare soil, 54
Allometric equation, 98, 102, 105
  ratios, 101
Apoplasmic water, 137
Assimilate allocation, 104
  coefficients, 110
  partitioning, 97
  supply, 97
Atmospheric stability, 59
Average irradiance, 49

"Biochemical" cycle, 112, 123, 129, 131
Biogeochemical cycle, 112, 114, 115, 116, 120, 122, 123
Biomass, accumulation, 19
  distribution, 99
  foliage, 127
  forest, 87
"Bottom-up" model, 166, 177

Bowen ratio, 45, 46, 47, 56, 62, 66, 176
Boundary layer, conductance, 75
  resistances, 50, 51
  thickness, 51
Buoyancy effects, 55

Calvin-Benson cycle, 70
Carbon balance of, trees, 69
  forest stands, 87
Carboxylase enzymes, 85
Carboxylation efficiency, 72, 81, 84
Canopy assimilation, 89
  conductance, 59, 63
  density, 67
  dry matter production, 88
  dynamics, 32
  energy balance, 43
  microclimates, 54, 61
  photosynthesis, 89
  resistance, 59
  storage capacity, 153
Capacitance, 144
Cell turgor, 160
Communities, 1
Consequences of water stress, 156
CO_2 assimilation rates, 71, 72
CO_2 efflux, 93
Crown leaching, 114

Darcey's Law, 146
Dark reactions, 70
Demand function, 72
Densities of wet wood, 143
Density of solid material in wood, 144
Dew point, 25
Diameter at breast height, 8, 98
Diffuse radiation, 22, 49
Direct beam irradiance, 22

Dimensionless drag coefficient, 59
Drag force, 56
Drainage, 134, 152
Dry matter increment, 117
  partitioning, 88, 95
  production, 67, 69, 90, 117, 172
  production, total annual, 92
Drought, 158

Ecosystem, 116
Eddies, 55
Effect of fertilization, 19
Effective precipitation, 152
  root length, 148
  root absorbing surface, 148
Electron carriers, 70
  transport, 70
Element concentrations, 125
Energy interception, 67
  conversion efficiency, 173
  partitioning, 57
Environmental limitations, 173
Errors associated with measurements, 8
Eucalyptus globoidea, 136, 137
  E. globulus, 19
  E. maculata, 33, 34
  E. marginata, 138
  E. obliqua, 40, 124
  E. pauciflora, 72, 74
  E. regnans, 18, 19, 40
Evaporation of intercepted water, 152, 154
Exchange coefficients, 57, 60
  processes between leaves and air, 50
  processes in canopies, 60
Extinction coefficient, 48, 49

Fagus sylvatica, 36
Fick's law of diffusion, 50
Fine roots, 115, 116
Fine root biomass, 103
  dynamics, 108
  production, 108
  turnover, 108, 109
Floral development, 11
Flow resistance, 150
  through trees, 162
Foliage area density, 34
  distribution, 35, 93

Foliage biomass, 127
  expansion, 158
  litterfall, 125
  mass, 33, 40
  nutrient concentrations, 117
  temperatures, 61
  water potential, 157
Forest biomass, 87
  catchments, 151
  ecosystems, 120
  productivity, 69, 89, 90
Forced convection, 25
Frankia, 120
Friction velocity, 56
Full canopy, 37

Gas constant, 135
Geochemical cycles, 112, 113, 114

Heartwood, 122
Heat, latent, 44, 50, 52, 78
  sensible, 45, 50, 52
  stored, 45
  specific, 45
  units, 11
Hierarchical processes, 7
Hofler diagram, 135
Humidity mixing ratio, 26
Hydrological balance, 151
Hydrologic equation, 151

Imminent competition, 42
Incident energy, 66
Infiltration capacity, 152
Interception losses, 152, 153
Intercepted energy, 117
  water, 61
Intercellular (CO_2) concentrations, 72
Irradiance, 22, 74, 81
  direct beam, 22
  diffuse, 22
  intensity distributions, 49
  total, 49

Laminar layer, 51
Leaf area, 37, 41, 88
  density, 53

index, 32, 34, 35, 36, 89
index, shaded, 48
index, sunlit, 48
profiles, 34
specific, 40
Leaf boundary layers, 51, 55
layer resistance, 52
emergence rates, 11
Leaf conductance, 74
dark respiration, 84
dynamics, 37, 38
energy balance, 50, 62
growth, 84
longevity, 129
nutritional status, 170
photosynthesis, 70, 71, 72, 73
protein, 71
shedding, 158
time course simulation, 161
transpiration, 53
volume, 145
water potential, 78, 141
water status, 16
Leaching, 114
Level of organization, 2, 13
molecular and biochemical processes, 13
response times, 13
Light harvesting chlorophyll, 71
interception models, 88
saturated assimilation rate, A_{max}, 84, 85
saturated photosynthesis, 83
Litter, 111
Logarithmic (wind) profile, 56
Long-wave radiation, 43

Management, 4, 165
Maximum leaf conductance, 77, 78
radiation intensity, 21
Mechanistic growth models, 166
Mesophyll conductance, 80
Microclimate, 31, 42
Micrometeorology, 42
Mineral nutrient concentration, 117, 123
Mineralization, 116, 122
Minimum infiltration area, 155, 156
Mobilizing ability, 96

Modelling of processes at different levels, 15
Model, 2, 4, 167
"bottom-up", 166, 177
complex, 170
diurnal course of canopy photosynthesis, 89
empirical, 2, 168
empirical, leaf photosynthesis, 79, 80, 81
explanatory, 168
growth as a function of nutrition, a model, 126
mechanistic, 3, 67, 169
non-transportable, 165
physiological, 10, 167, 171
sub-, 168, 177
testing, 170, 171
"top-down", 166, 170, 172, 173, 176, 177, 178
water movement through trees, 143
Modifiers, 173
nutrition, 175
temperature, 174
water, 175
Molecular and biochemical processes, 13
diffusion, 51
Momentum absorption, 55
exchange coefficient, 56
flux, drag force, 56
transfer, 55
Multivariate analysis, 9, 174
Mycorrhizae, 122
Mycorrhizal fungi, 119

N-mineralization, 115, 116, 130
N-fixation, 120
Net photosynthesis, 18
Net radiation, 43, 50, 62, 149
Nitrogen, 111, 114, 115, 116, 117, 121
cations, 114
concentration, 128
productivity, 127
Nothofagus solandri, 16, 17
Null hypothesis, 9, 10
Nutrient concentrations, 116, 124, 126
cycling, 111, 112, 113
dynamics, 110

Nutrient concentrations—*cont.*
  flux density, 128, 129
  losses, 126
  losses, fire, 120
  losses, harvesting, 121
  movement, 111
  reserves, 111
  status, 130
  supplying ability, 126
  translocation, 118
  turnover, 115
  uptake, 112, 116, 117, 118, 119
  utilization efficiency, 130

Ohm's law analogue, 50
Organizational levels, 17
Organic matter, 115
Organic matter, decomposition, 115, 121
Osmotic adjustment, 157
Osmotic potential, 135, 138
Osmotic pressure, 135

Partitioning absorbed energy, 57
  coefficients, 92, 96, 98, 123
  heritability, 110
  ratios, 93
Pascal, 25
Penman-Monteith equation, 63, 64, 65
  aerodynamic term, 150
Phosphorus, 117, 121, 124
  translocation, 132
Photon-flux density, 14, 18, 80, 88, 89
Photosystems, 71
Photosynthesis, *see also* Light saturated photosynthetic assimilation rate, 14, 70, 71, 82, 129, 134, 172
Photosynthetic characteristics of the foliage, 117
  empirical model, 79, 80, 81
  properties of leaves, 79
  rates, 157
  surface, 129
Photosynthetically active radiation, 20, 21
Physiological modelling, 10
Picea abies, 127
  P. sitchensis, 13, 63, 76, 77, 84, 99, 119

Pinus contorta, 102
  P. densiflora, 40
  P. nigra, 124, 127
  P. ponderosa, 54
  P. radiata, 18, 19, 43, 63, 98, 111, 121, 124, 138, 174
  P. resinosa, 97, 127
  P. sylvestris, 43, 63, 101, 102, 127, 150
  P. taeda, 124
Plant-available (nutrients), 118
Plant–water relations, 133
  status, 134
Population of trees, 39, 41, 176
Porometers, 63
Potential gradients, 103
  matric, 135, 138
  osmotic, 135, 137, 138
  pre-dawn water, 156, 159
  soil water, 138
  time course, of leaf water, 160
  tissue water, 134
  total water, 135
  turgor, 135
  water, 135
  xylem water, 143
Precipitation, 27
Prediction of forest growth, 178
Pressure–volume curves, 136, 137
Pseudotsuga menziesii, 35, 62, 63, 76, 127

Quantum flux (photon-flux density), *see also* Photon-flux density, 18
Quantum yield, 82, 83

Radiant energy, 32, 52, 172
  absorbed, 66
Radiation, direct beam, 47
  diffuse, 47
  interception non-continuous canopies, 49
  penetration into plant stands, 47, 48, 49
Radiating surface, 43
Radiative temperature, 43
Rainfall, amounts, 27
  distribution, 27, 28

Index

effectiveness, 28
frequency, 28
redistribution, 154
Rainsplash, 154
Range of states and processes, 14
Rates and states, 7
Rate measurement, 8
Reducing power, 70
Reflection coefficient, 43
Reflectivity of surfaces, 43
Relative humidity, 25
  leaf growth rate, 128
  water content, 135, 136, 137
Resistance–capacitance model, 10
Resistance to water flow through the soil, 141
  to the flow of water through trees, 36
Respiration, 88, 93, 94, 98, 172
  losses, 95
  maintenance component, 93
  stem, 94
Response times, 23
Retranslocation, 119
Rhizobia, 120
Ribulose–bisphosphate (RuBP), 71
Root length distribution, 138
  system, 103, 104, 105, 106, 118
  structural, 109
  temperatures, 25
  turnover, 105
Root fine, 106, 115
  biomass, 103
  growth, total, 106
  length, 147
  production, 105, 106, 107
  production, biomass, 107
  resistance to water uptake, 109
  turnover, 103, 105, 109, 110
Root–length/root–diameter relationship, 103
Root–soil contact, 147, 148
Rooting zones, 177
  volume, 137
Ratios, branch weight/stem weight, 101
  stem weight/total weight, 101
  weight/total weight, 101
Roughness element density, 57, 58
RuBP carboxylase, 70
Run-off, 134, 152

Sapwood, cross-sectional area, 33, 40
  volume, 143
Saturation water vapour pressure, 25
Self-thinning, 39, 42
Shearing stress, 56
Shoot elongation rates, 18, 19
Short-wave radiant energy, 20
  radiation, 43, 49, 50
Sink activity, 97
  strength, 96
Soil aeration, 121
  bulk density, 139
  heterogeneity, 152
  temperatures, 24, 121
  wetness, 139
Soil hydraulic characteristics, 130, 138
  conductivity, 139, 140, 152
  diffusivity, 140
  properties, 147
  saturated, 152
Soil–root resistance, 146
Soil–water characteristics, 177
  balance, 134, 159, 160
  content, 138
  potential in the root zone, 139
Solar zenith angle, 48
Spatial variation, 159
Specific activity, 96, 108
Specific leaf area, 40
Spherical leaf angle distribution, 48
Stand evaporation, 134
  structure, 31, 41, 67
  transpiration, 67
  water balance, 66
Standing biomass, 87
Statistical techniques, 9
Stem cross-sectional area, 146
  diameter, 8, 33, 40
  elongation rates, 11
  flow, 153, 155
  mass, 98
  respiration, 94, 95
Stomatal aperture, 50, 51, 73
  conductance, 51, 63, 69, 73, 75, 76, 77, 79, 84, 85, 89
  resistance, 50, 51
  response to humidity, 78
Storage tissue, 143
Symplasmic water, 136, 137
Surface erosion, 114

Sustained yield, 122
Sunflecks, 49
Sunshine hours, 22, 62
Sunlit leaf area index, 48
Symbiotic fixation, 115
  non-symbiotic fixation, 121
  relationships, 120

Tectona grandis, 77
Temperature, effects of, 10, 11
  maximum and minimum 24
  optimum range, 174
  profile, 54
Theoretical quantum requirement, 82
Thermal time, 11, 12
Thermal time model, 13
Thinning, 67
Throughfall, 153, 154
Time course of leaf water potential,
  model, 161, 162
Time series data, biomass, 100, 101
Tissue death rates, 98
  water potentials, 36
"Top-down", model, 166, 173, 176, 178
Total annual dry matter production, 92
  daily energy income, 21
  irradiance, 49
Translocation to, 97
Transpiration, 26, 61, 63, 150
  by a forest, 62
  rate, 62, 133, 134, 141, 162
Tree biomass, 98
  growth rate, 116
  population, 90, 116
  respiration, 95
Turbulent diffusion, 51, 55
  exchange, 65

transfer, 32, 53
transfer processes, 55
Turgor maintenance, 157
Tsuga heterophylla, 99

Understorey vegetation, 116

Vapour pressure profiles, 54
Vapour pressure, 25
  deficit, 26, 63, 74, 75, 78, 149
  gradients, 26
von Karman's constant, 56
Visible radiation, 48

Water apoplasmic, 137
  intercepted, 61
  modifier, 175
  movement through soil, 139
  movement through trees, 141
  relations, 66
  stress, 70, 75, 78, 157
  stress, consequences of, 157
  symplasmic, 136, 137
  use efficiency, 158
Weather conditions, 20
Weathering, 114
Wilting point, 151, 176
Wind speed, 27, 52, 62

Xylem water potential, 143
  conducting tissue, 144

Zero-plane displacement, 56, 57